MICROBIOLOGY OF FROZEN FOODS

MICROBIOLOGY OF FROZEN FOODS

Edited by

R. K. ROBINSON

M.A., D.Phil.

Department of Food Science, University of Reading, UK

ELSEVIER APPLIED SCIENCE PUBLISHERS

LONDON and NEW YORK

ELSEVIER APPLIED SCIENCE PUBLISHERS LTD
Crown House, Linton Road, Barking, Essex IG11 8JU, England

Sole Distributor in the USA and Canada
ELSEVIER SCIENCE PUBLISHING CO., INC.
52 Vanderbilt Avenue, New York, NY 10017, USA

British Library Cataloguing in Publication Data

Microbiology of frozen foods.
1. Food, Frozen   2. Microbiology
I. Robinson, R. K.
664'.02853   TP493.5

ISBN 0-85334-335-7

WITH 52 ILLUSTRATIONS AND 29 TABLES

© ELSEVIER APPLIED SCIENCE PUBLISHERS LTD 1985

The selection and presentation of material and the opinions expressed in this publication
are the sole responsibility of the authors concerned

Printed in Northern Ireland by The Universities Press (Belfast), Ltd

Preface

It is widely assumed that frozen foods do not pose a microbiological threat to consumers and, in general, this confidence is justified. Nevertheless, success does depend on the fact that the initial foodstuffs are, prior to freezing, of a sound hygienic quality and, furthermore, that the thawing/cooking operations reflect the properties of the frozen product. The achievement of these twin aims, without adverse effect on the organoleptic characteristics of the food, has been the essential aim of the manufacturers of frozen foods, and the increased usage of home freezers is a tribute to their success.

Some of the problems associated with the freezing of different commodities, together with the techniques for overcoming them, are discussed in this book, and if this knowledge will help to further the safety of frozen foods, then its collation will have been fully justified.

R. K. ROBINSON

v

Contents

List of Contributors

Mr M. F. G. BOAST
3 Farrar Top, Markyate, St. Albans, Hertfordshire AL3 8NA, UK.

Professor L. BOEGH-SOERENSEN
Department of Food Preservation, The Royal Veterinary and Agricultural University, Howitzvej 13, DK-2000 København F, Denmark.

Dr R. DAVIES
National College of Food Technology, Food Studies Building, University of Reading, Whiteknights, Reading RG6 2AH, UK.

Dr C. O. GILL
The Meat Research Institute of New Zealand, PO Box 617, Hamilton, New Zealand.

Mr L. P. HALL
Campden Food Preservation Research Association, Chipping Campden, Gloucestershire GL55 6LD, UK.

Dr M. JUL
Department of Food Preservation, The Royal Veterinary and Agricultural University, Howitzveg 13, DK-2000 København F, Denmark.

Miss E. C. LAMPRECHT
Fishing Industry Research Institute, University of Cape Town, Private Bag, Rondebosch 7700, South Africa.

Mr P. D. Lowry
The Meat Research Institute of New Zealand, PO Box 617, Hamilton, New Zealand.

Dr A. Obafemi
Department of Microbiology, University of Port Harcourt, Nigeria.

Dr R. K. Robinson
Department of Food Science, University of Reading, London Road, Reading RG1 5AQ, UK.

Dr J. Rothwell
Department of Food Science, University of Reading, London Road, Reading RG1 5AQ, UK.

Dr C. K. Simmonds
Fishing Industry Research Institute, University of Cape Town, Private Bag, Rondebosch 7700, South Africa.

Miss C. A. White
Campden Food Preservation Research Association, Chipping Campden, Gloucestershire GL55 6LD, UK.

Chapter 1

The Technology of Freezing

M. F. G. Boast

Markyate, St. Albans, UK

INTRODUCTION

Much of our food cannot be stored for any length of time without loss of quality, for, within most foodstuffs, microbiological, chemical, biochemical and physical processes occur causing deterioration. Refrigeration can play a vital role in controlling these processes, because all of them are retarded by lowered temperature.

The first record of this effect was made by Tellier in 1872, when he reported that meat stored between −2°C and +3°C retained its fresh quality, and in 1892, Grassman observed 'no harmful change nor loss of nutritive value' in pork and beef stored at temperatures from −2°C to −4°C. Experiments in refrigeration were also being carried out in Australia, and attention was directed towards an artificial process which could 'produce cold', and apply this to preserving a carcass of meat for a lengthy period of time.

One of the first practical demonstrations of this effect was on the 5th of February 1882, when the first cargo of frozen meat, consisting of 4311 sheep and 598 lambs, was despatched from New Zealand. The cargo arrived in London in excellent condition (Cameron, 1904), and the use of refrigeration for food storage had begun in earnest. Techniques for lowering temperature have, of course, advanced a great deal since these early days, but as shown below, the basic principles have remained unchanged by time.

1

COMPRESSION–EXPANSION SYSTEMS

Compression–expansion refrigeration systems provide the basis for the majority of freezing operations, and the principle of this system is shown in Fig. 1. The refrigerant vapour is circulated by the prime mover in the form of a compressor, which draws a low pressure, low temperature vapour from the evaporator, and compresses it to a higher pressure and temperature. In the condenser, heat is removed from this vapour—initially the superheat and then the latent heat—causing the vapour to condense at constant temperature and pressure into a liquid; additional cooling will further reduce the liquid temperature (sub-cooling). The condenser heat is given up to the medium (air or water) surrounding the unit.

The high-pressure, liquid refrigerant then passes through the control device where the pressure is reduced, and some of the liquid evaporated (flash gas) in order to cool the balance of the liquid to the evaporating temperature. In the evaporator, the refrigerant absorbs heat taken from the product, and returns to the vapour state. A small amount of superheat is given to this vapour (3°C to 10°C) to ensure that only vapour returns to the compressor, so preventing possible damage to the pump. The heat to be removed from the condenser (Q_C)

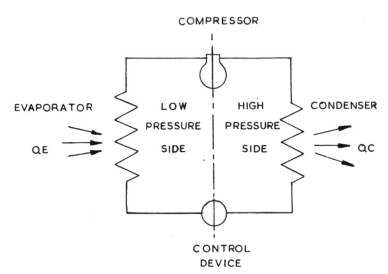

Fig. 1. The compression–expansion refrigeration system.

will exceed the heat (Q_E) taken in at the evaporator because of the work done on the refrigerant in passing through the compressor.

Refrigerants

Closed loop recirculation is employed in a wide range of freezing and storage systems, and the compounds used as refrigerants have been given International R (Refrigerant) Numbers (Boast, 1982). (See Table I.)

Refrigerants R717 (ammonia), R12, R22 and R502 are commonly used in refrigerating systems for frozen foods. Ammonia is the sole survivor of the classical refrigerants, the others, such as air or carbon dioxide, having been superseded because of the low thermodynamic efficiencies associated with their use. The other three refrigerants commonly used are compounds of chlorine, fluorine and carbon belonging to a range of synthetic compounds known as chloro-fluoro-carbons (CFCs).

Ammonia is a toxic, strongly reactive compound with a great affinity for water, whereas CFCs are generally non-toxic, stable compounds with very little ability to absorb water. Nevertheless, ammonia has remained in use because of its excellent thermodynamic properties, and its ability to give high rates of heat transfer at low pressure drops. The use of ammonia in refrigerating systems is well understood, and there is a good safety record. Further advantages of ammonia include its strong smell, high latent heat, low viscosity, high acoustic velocity and low liquid density. In addition, because ammonia is toxic, such systems tend to be robustly constructed, and this strength is in itself

TABLE I
Typical refrigerants used in industrial and domestic freezing and storage

Refrigerant	Chemical formula	Boiling temperature at 1 bar pressure (°C)	Enthalpy of latent heat (kJ kg^{-1}) at $-20°C$	Typical application
R12	CCl_2F_2	-30	160	Commercial
R22	$CHClF_2$	-41	220	Air conditioning Low temperature
R502	$CHClF$ $CClF_2CF_2$	-46	162	Low temperature
R717	NH_3	-34	1324	Industrial

desirable under operating conditions. The disadvantages of ammonia include its toxicity, its corrosive effect on copper, zinc and most plastics, its high index of compression and its immiscibility with lubricating oil.

CFC refrigerants were developed because stable, non-toxic refrigerants were sought for small, fully automatic refrigerators. The advantages of CFC refrigerants include chemical stability, miscibility with oil, low index of compression, non-conductance of electricity, relatively good thermodynamic properties and very low toxicity. The disadvantages include very low latent heats, generally low acoustic velocities, absence of smell and very low moisture solubility. At one time, price was also a serious obstacle to the use of CFCs, but prices are now such that the cost of ammonia or a CFC refrigerant forms a very small part of the cost of the installed system.

Refrigerant Control and Circulation

Refrigerating systems of the type under consideration require a control device to meter the high pressure, liquefied refrigerant into the low pressure side of the system where it can absorb heat by evaporating. Provision must also be made to circulate the refrigerant through the evaporator.

Systems may be divided into 'once through' types in which the refrigerant is injected directly into the evaporator and then drawn away to the compressor, and recirculatory systems in which the low pressure refrigerant is recirculated through the evaporator either by natural circulation, or by pump and overfeed systems; in the latter, the refrigerant is injected directly into the evaporator prior to extraction in a wet state and superheating elsewhere by heat exchange.

The typical metering device for the 'once through' system is the thermostatic expansion valve, which is actuated by the superheat of the refrigerant leaving the evaporator. Thermostatic expansion valves can be employed with CFC refrigerants or ammonia, but their use with ammonia is less satisfactory because: (a) there is a danger of flooding owing to the small volume of liquid refrigerant to be metered; (b) the risk of seat erosion in the valve; (c) poor contact between thermal bulb and refrigerant owing to the use of heavy steel pipe; and (d) the greater effect of small droplets of liquid which inevitably pass the expansion valve phial unless very large superheats are used.

The advantages of thermostatic expansion valves include their small size, their cheapness, their ability to meter refrigerant to many evaporators in parallel, their capacity to adjust to varying loads and their ability, when working correctly, to prevent flooding of the refrigerating compressor without the need for large separating vessels or superheaters in the compressor suction line. Their disadvantages include their fundamental need for a considerable degree of superheat to be produced in the evaporator, and the fact that they do not 'fail safe' when the seat becomes worn, or an optimistic machinery minder adjusts the superheating setting to a level at which the control is not stable.

Recirculatory systems may be fed by high pressure float-type controls. The high pressure device is sensitive to the level of liquid in the high pressure side of the system, and operates to keep the high pressure side drained to that level. The advantages of the device are that the control adjusts smoothly to cope with varying loads, and that it is not necessary to provide a storage vessel for high pressure liquid. The disadvantages of the device are that only one low pressure system can be fed from it, and that liquid sub-cooling in two-stage plants is made complicated.

The alternative, the low pressure float control, is sensitive to the liquid level in the low pressure side of the plant, and has the advantage that multiple circuits at varying pressures may be fed using one control per circuit. The disadvantages of the device, however, are that storage vessels for liquid are required both on the high pressure and low pressure sides of the system, and, furthermore, its use tends to cause wide fluctuations in fluid level, particularly when the refrigerant circuits in the evaporator are long. This variation occurs because the additional flash gas, liberated when the device opens on a fall of liquid level, raises the pressure (and tends to suppress boiling) in the evaporator, which consequently can contain more liquid than it would if the boiling rate was steady. When eventually the control level rises and the supply of flash gas is reduced by the closing of the valve, the consequent depression in evaporating temperature causes increased boiling in the evaporator which, in turn, vents some of its contents to the low pressure storage vessel causing a significant rise in level. This effect can be mitigated by injecting the flash gas into the down leg of natural circulation systems, but it is difficult to improve the performance of pump circulated systems in a comparable way.

The overfeed system is a special case of the 'once through' approach, in which the evaporator is fed in such a way that the whole of its

surface is used to evaporate refrigerant, and the vapour is drawn away
from the evaporator in a wet state. The wet vapour is separated into its
wet and dry fractions in a suitably sized vessel in the return line to the
compressor, and heat exchangers are used to vaporise the liquid
fraction, and also to superheat the vapour fraction. An overfeed
system may be controlled by high pressure float devices where the
system is a simple one with few evaporators, or it may be controlled by
probes which sense the degree of wetness of the vapour leaving each
cooler and adjust the flow of refrigerant accordingly. Overfeed systems
are well suited to CFC refrigerants which have large mass flows per
unit of refrigeration, and good miscibility with lubricating oil.
Ammonia would be much more difficult to use in an overfeed system,
because the amount of overfeed which could be arranged would be
very much less.

Overfeed systems may be applied to single- or two-stage refrigerated
systems, and it is the use of heat exchangers to improve thermodynamic efficiency, together with the elimination of refrigerant liquid
pumps and the reduced need for intercooling, which allows the design
of a two-stage CFC overfeed system with an efficiency similar to that of
a two-stage ammonia system (Forbes-Pearson, 1983).

FREEZING

The process of freezing requires the controlled removal of heat from
the product, at a steady uniform rate, until the heat remaining in the
product is equal to its equilibrium heat after stabilisation at storage
temperature. Thus, if the long-term storage temperature is −30°C,
then the outer surfaces of the product may have to be reduced to
−35°C or −37°C to ensure an equilibrium temperature of −30°C.
Removal of heat from the product after transferring into storage is not
recommended, since this will, in effect, change the rate of freezing and
not aid the quality. It is better, therefore, to over cool a product in the
freezer and then move it to storage, even though the store may be a
few degrees warmer.

The freezing process results in ice crystals being formed inside the
product, and the size, or change in size, of these crystals will determine
the quality of the product on thawing. The increase in diameter of the
ice crystals takes place mostly during the first month of storage, and it
is important to note that the growth of the crystals is greater at higher

and/or uneven temperatures. The rate of growth, expressed as a percentage of the original size, is approximately the same for all sizes of crystal. Thus, if the original crystals were less than 100 μm in diameter, they will grow to, say, 120 μm in storage, whilst large crystals of, say, 400 μm, will increase to 500 μm in the same time. Low and constant temperatures retard the growth of ice crystals.

The ultimate test of the freezing process is whether frozen products resume their original quality upon defrosting, and, in this context, speed of freezing can be critical.

Freezing Time

The 'freezing time' refers to the time required for the entire temperature change, and it is important to understand the relationship between the capacity of a freezing system and the freezing time of the product. Each product has an associated freezing time, which depends both on product composition and dimensions, and the effectiveness or efficiency of the freezing system. Within a particular product group, product size significantly influences the freezing time, and as all of the heat has to leave the product through its surface, the relationship between surface and weight is of great importance; as volume and weight are proportional, the two terms are interchangeable in this context. Freezing time is inversely proportional to the specific surface for small particles where temperature gradients within the product can be neglected, but for a package, the temperature gradient is critical, and the freezing time will be approximately proportional to the thickness of the package. Thus, the capacity of a freezer will be reduced by 50 per cent if the thickness of a product is doubled.

Freezing times are generally determined in the industry by experience, or by technicians working with specialised laboratory test freezers.

Packaging

The packaging used for frozen foods should be of good quality, as well as odour and taint-free, in order to prevent contamination of the product by exposure or handling. Thus, the product should be totally enclosed, and the container should not be capable of being opened without recognisable damage. It should also protect the product against normal transit and storage hazards, and inhibit dehydration by incorporating a moisture vapour barrier. All outer packaging should carry clear product identification, and should be coded so that stock rotation can be carried out. The inner pack or carton should be marked with an appropriate identification code which should enable the producer concerned to establish:

1. the date of production (final date of packing);
2. the location of the producing factory;
3. easy reference to daily production records and other circumstance of manufacture.

The information referred to in points 1 and 2 above, together with other relevant production details, should be available in relation to any individual packet, and hence accessible to any retailer, caterer or enforcement officer having reasonable cause to require such information in relation to the quality of the product. The inner pack or carton should also carry details of recommended storage times, including reference to star marking as set out in British Standard 3739, and, as applicable, thawing times and cooking instructions.

The freezing process can be carried out with the product packed or unpacked, but freezing times will be quicker if no wrapping is around the product; any packaging acts as an insulator and reduces heat transfer. When a product is wrapped, the inner layer must have excellent contact with the actual food, and all air must be removed if reasonable cooling rates are to be obtained. If an additional outer packing is used, the individual items should be arranged to make a regular shape, and hence establish good contact with the surrounding material.

Cooked foods should be chilled to around 10°C or lower before entering the freezer, and this temperature should be checked at the core. Thus, the freezing process takes place from the outside and works towards the centre, so that the moment the outer surfaces are frozen, they act as a barrier which slows the rate of heat transfer from the core. Hence, if a hot product is placed in a freezer, the time taken to freeze it through to the centre can be excessive, resulting in poor quality when thawed. Similarly, food cooked in a domestic kitchen should first be allowed to cool close to ambient, and then be transferred to the refrigerator to chill down before being placed in the freezer. The use of an immersion thermometer to check core temperatures prior to freezing is highly recommended.

Weight Loss

One essential factor to consider when choosing freezing equipment is the weight loss that occurs during freezing, for the costs saved through efficiency can be about the same as the operating cost of the freezer. This dictum applies even with inexpensive products like peas, and is much more significant for expensive products like seafoods.

Weight loss during freezing can be caused by mechanical losses, downgrading, and dehydration. Mechanical losses refer to products dropping to the floor, sticking to conveyor belts, or simply the dripping of juice, but a modern freezer should have almost no losses in this category. Downgrading losses refer to breakages and similar occurrences which render the product unsaleable at the top quality price. Dehydration losses will be present in any freezing system, for although the evaporation of water vapour from unpacked products during freezing cannot be seen, it soon becomes evident as frost built-up on evaporator surfaces. This frost can also be caused by excessive infiltration of warm, moist air into the freezer. Still air inside a diffusion-tight carton often creates larger dehydration losses than occur with unpacked products, because with no circulation of air within the package, heat transfer is poor. The result is evaporation of moisture with the frost staying inside of the carton.

A poorly-designed freezing tunnel operates with dehydration losses of 3 to 4 per cent, while a well-designed tunnel can operate with losses of 0·5 to 1·5 per cent. Cryogenic freezing systems normally operate with a dehydration loss of about 0·5 to 1·25 per cent; this loss occurs when the gas is circulated over the product at the 'infeed' end of the freezer. The liquid refrigerant method operates with the lowest dehydration loss of any known method—generally in the region of 0·1 to 0·25 per cent. Dehydration is low because the product entering the freezer is in immediate contact with the boiling liquid refrigerant, and hence is instantly 'crust-frozen'; this prevents the vaporisation of moisture from the product. With any freezing system, rapidly crust-freezing the product is imperative if dehydration losses are to be minimised.

FREEZING SYSTEMS

The removal of heat from a product will require a refrigeration freezing system, and various types are used depending on:

    (a) the shape and size of the individual items of food;

(b) the total quantity of product to be handled per batch; and
(c) the availability of material, i.e. seasonal or year-round.

Many food items can be characterised by their relatively consistent particle size. These foods include not only peas, potato chips, whole kernel corn and cut green beans, but also meatballs, shrimps and fish fillets, and, for this collection of products, it is desirable to have every particle frozen individually so that there are no clumps or blocks. This approach is known as the IQF (Individually Quick Frozen) system, a technique that gives better heat transfer, and improves the overall efficiency of the process. Thus heat transfer from a product particle is many times more efficient if the particle is treated individually rather than in the aggregate, for while the ratio of surface to mass is increased, the insulating effect of a collection of particles is reduced. These products would be suited, therefore, to a fluidised-bed freezer, in which the items are held in suspension by an upward blast of cold air.

Pastes, thick liquids and thin, flat items, such as fish fillets, can be frozen on a rotating drum freezer with good results and minimum dehydration, but large items should be frozen individually. Whole chickens or poultry in pieces can be frozen in an air-blast freezer of the stationary tunnel or 'push-through' type, or in an automatic-feed freezing tunnel. Products which have been bulked together, such as fish, which is often compacted into 25 kg blocks very soon after harvesting, are best frozen in a plate freezer—either horizontal or vertical. Plate freezers can also be used for paste or slurry-type foods, but the disadvantage of the plate freezer is the need to defrost the plates to release the product. This process may defrost, in turn, the outer surface of the product, so reducing slightly the prospect of the food retaining acceptable quality in the long-term.

However, irrespective of the nature of the foodstuff, the basic intention of the process is the same, namely to achieve a rapid removal of heat, and this can be accomplished in one of three ways:

1. Contact with cold air or surface produced by a closed-loop recirculation system of refrigeration;
2. The use of a cryogenic total loss system;
3. Immersion in a liquid refrigerant,

and the application of these techniques is discussed in the following pages.

Systems Involving a Standard Refrigeration Unit

Storage room

A storage room should not be considered as freezing equipment, even though it is sometimes used for this purpose, for this approach has so many disadvantages that it should be used only in exceptional cases. For one thing, the freezing is so slow that the quality of almost all food suffers, and if existing products are already being stored in the room, their quality is further jeopardised because flavours may be transferred from the warm products yet to be frozen. Because a storage room is not designed for freezing, the cooling coils may frost up so quickly that the total refrigeration capacity is reduced below the level required to maintain proper storage temperature. The temperature of the already frozen products may then rise considerably, to the detriment of their quality. In general, therefore, freezing in a storage room gives rise to poor quality foods, and incidently, leads to higher costs than using an in-line freezer specially designed for the application.

Air-blast systems

Because air is the most common freezing medium, equipment designs vary widely, and among the systems employed are: blast rooms, stationary tunnels, push-through tunnels, automatic tunnels, belt freezers and fluidised-bed freezers.

Blast room

A blast-freezing room consists of an insulated room equipped with a greater than normal number of forced-air cooling units. The air coolers are equipped with fans which create turbulence in the air. Products are laid on trays, which are loaded into racks, and the tier spacing in the rack is designed to provide an air space between the trays of at least 50% of the product thickness. Because there is poor air-circulation control, the resulting heat transfer at the product/air interface is low. The blast room offers, therefore, limited freezing potential.

Stationary freezing tunnel

The stationary freezing tunnel is the simplest type of modern freezer, and is designed to produce satisfactory results with the majority of products. It consists of an insulated enclosure equipped with refrigeration coils and fans that circulate the air in a controlled pattern

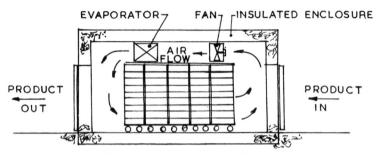

Fig. 2. Push-through rack freezer.

over the product. The characteristics of this air-flow influence the freezing rate and the associated loss in product weight; a typical installation is shown in Fig. 2. Products are placed on trays which are then placed into a rack. The racks are arranged to provide an air space between each level of trays, and are moved in and out of the tunnel manually. The human element becomes important when situating the racks inside the tunnel, for the key to a more effective freezing process is ensuring that the flow of air cannot by-pass the trays of product. Practically all foods can be frozen in a stationary freezing tunnel. Whole, sliced, or diced vegetables may be frozen in cartons, or unpacked in a layer (30 to 40 mm deep) on trays, while spinach, broccoli, meat patties, fish fillets and prepared foods are often frozen in packages in this type of equipment; by using different rack designs, thick packages and whole meat carcasses can also be frozen. The capacity of the system depends on product thickness and composition, as well as on the presence or absence of packaging, but the flexibility of this type of freezer makes it especially suitable during the initial stages of development of a new frozen food. However, it does require a heavy outlay of manpower, and can give rise to considerable weight losses if improperly used.

Push-through tunnel
A degree of mechanisation is achieved in a 'push-through' tunnel in that the racks are fitted with casters or wheels, and are usually moved on rails by a mechanically powered pushing mechanism, products with different freezing times having separate racks and rails. This type of freezer has the same basic advantages and disadvantages as the stationary tunnel.

Automatic freezing tunnel

The demand for automatic operation has led to a great variety of freezing tunnels with sophisticated mechanisation, and among these freezers are: the sliding tray freezer, the travelling tray freezer, the carrier freezer and the spiral freezer.

Basically, the sliding tray freezer consists of one great rack that accommodates many big trays on each tier, and at one end of the system, an elevating mechanism lifts the in-coming trays to the top tier. Here, they are pushed inward, forcing all the other trays in the tier to advance one step. The tray at the far end is pushed onto an elevator, lowered one tier, and then pushed in the return direction. Thus, in every odd tier, the trays advance, and on every even tier, they return. For each tray that enters, all trays advance one step.

In another version of the automatic freezing tunnel, the trays move on only one tier at a time, which gives almost the same result, and this version is sometimes equipped with a plate freezer arrangement on the first 20 to 30 per cent of each tier. This modification eliminates the bulging of packages, but requires more space. The mobile sections of the system are usually powered hydraulically, and outside the freezer enclosure, automatic loading and unloading of the trays is also provided. It is important to note, however, that each tray is exposed to considerable mechanical stress which limits both width and length, and hence this type of freezer is only suitable for handling intermediate size packages at moderate capacity.

The carrier freezer may be regarded as two push-through tunnels on top of each other. In the top section, the carriers are pushed forward, and in the lower section, they are returned; in both ends, there are elevating mechanisms. Each carrier is similar to a bookcase, and when it is at the loading end of the freezer, frozen product is pushed from the carrier, one shelf at a time, onto a discharge conveyor. As the carrier is indexed upward, unfrozen product is transferred from an infeed conveyor onto the just emptied shelf. The carriers may be designed for almost any pitch between shelves, and for variable lengths and widths.

Spiral belt freezers

Because the belt is continuous, internal transfer points are eliminated, and the product is transferred only at the infeed and outfeed ends of the freezer. Products are placed on the belt outside the freezer where the transfer can be monitored, and because the belt is continuous, the

Fig. 3. Spiral freezer for use with a mechanical refrigeration plant. (Courtesy of APV plc.)

product occupies the same area throughout the freezing process. Similarly, no significant product movement occurs with respect to the belt. By employing a single belt, a continuous cleaning process can be installed for those products that warrant it, while the flexibility of the belt allows for more than one infeed and outfeed. Furthermore, the relative locations of the infeed and outfeed can be arranged to suit the requirements of a particular manufacturer.

Spiral belt freezers are suitable for unpackaged meat, fish and poultry products, including meat patties, meat balls, fish fillets, and cut chicken portions. The versatility of the spiral design also makes it suitable for packaged products such as prepared meals, and it is one of the few conventional freezing tunnels that can be used for freezing soft-formed products, such as raw meatballs or patties. (See Fig. 3.)

Fluidised-bed freezers
Fluidisation is defined as a method to keep solid particles floating in an upward-directed flow of a gas or a liquid. In freezing, fluidisation

occurs when particles of similar shape and size are subjected to an upward stream of low temperature air. At a certain air velocity, the particles will float in the air stream, each particle separated from the others but surrounded by air and free to move. In this state, the mass of particles will assume the properties of a liquid. The fluidisation principle, when applied to an in-line freezer, corresponds to the concept of a dam with a spillway. If the particles are retained in a cubical area that is designed with an open end (the outfeed end) slightly lower than the other three walls, product fed in at the infeed will displace product in the fluidised bed and cause a flow through the freezer bed. By utilising low temperature air to achieve the fluidisation, the product is frozen and simultaneously conveyed by the same air without the aid of a mechanical belt. Figure 4 shows the principle of the system, and Fig. 5 the product flow.

Using the fluidisation principle for freezing provides several advantages over the more conventional belt. All products, including sliced green beans, sliced carrots and sliced cucumbers, which tend to stick together are individually quick-frozen, and the process is especially effective and reliable when freezing wet products; a deep fluidised bed can accept products with high surface water contents. Another advantage of the fluidised-bed freezer is its complete independence from variations in product flow, for even when running at reduced capacity, the same, evenly distributed air pattern is maintained without channelling.

Through fluidisation and effective air/product contact, heat transfer rates higher than those considered normal for conventional air-blast

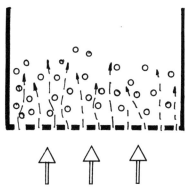

Fig. 4. Fluidisation principle.

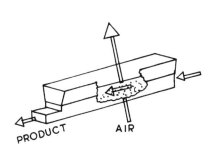

Fig. 5. Fluidised-bed freezer.

freezing tunnels are achieved. An indication of this efficiency can be seen in the physical dimensions of the freezer, which are generally one-third those of a comparable capacity belt freezer. The fluidised-bed freezer is suitable for vegetables, fruits and berries, as well as processed products like french fried potatoes, peeled cooked shrimps, diced meat or cooked meatballs. Products may vary from small uniform items up to 40 mm in size, to larger products up to 125 mm in length, and this tolerance allows maximum utilisation regardless of season. Stainless steel is employed for all surfaces in direct contact with foodstuffs, and an hourly throughput of between 2 and 5 tonnes would be anticipated. (ASHRAE, 1982).

Direct contact systems

Plate freezers

In a plate freezer, the product is firmly pressed between top and bottom metal plates. The refrigerant is circulated in channels housed inside the plates, and this ensures good heat transfer and relatively short freezing times if certain design and operating criteria are met. Thus, the plates should be flat and free of distortion, and the packages must be tightly filled with the product; it is also an advantage if the product itself is a good conductor. All of these factors have a positive influence on heat transfer by conductance, the primary heat removal mode, and, because the efficiency of heat transfer is gradually reduced with increasing thickness of the product, package thickness is often limited to a maximum of 50 mm.

Pressure from the plates has a secondary positive influence during the freezing process, in that it eliminates bulging in packages, a common occurrence in air-blast tunnels, and the packages discharge with straight sides that are within close tolerances.

The two main types of plate freezer are the horizontal and the vertical, and either type can be manual or automatic. Horizontal plate freezers generally have 15 to 20 plates, and the product is placed on metal trays at the end of the packaging line, loaded in a rack or trolley, and transported to the freezer. The trays are then manually loaded between the plates. A typical horizontal plate freezer is shown in Fig. 6.

Automatic operation of a horizontal plate freezer can be obtained by designing the entire battery of plates to move up and down in an elevating system. At the level of the loading conveyor, the plates are

Fig. 6. Typical horizontal plate freezer. (Courtesy of APV plc.)

separated, and packages that have accumulated on an infeed conveyor are pushed between the plates. This action has the simultaneous effect of discharging a row of frozen packages at the opposite end of the plates, and the cycle is repeated until all the frozen packages have been replaced. The plates are then closed, and indexed upwards.

The vertical plate freezer, developed specifically for freezing fish at sea, consists of a series of vertical freezing plates that form partitions in a container. Products are fed from the top, and the finished block of frozen product is discharged to either the side, the top, or the bottom; usually this operation is mechanised. Most whole, unpacked fish are frozen in this way, but fillets may also be handled in the vertical plate freezer. The thickness of the block can vary from 50 to 150 mm.

M. F. G. Boast

Drum freezer

The product to be frozen is placed on a stainless steel, refrigerated drum which rotates at a pre-set speed depending on the freezing time required; the product must be frozen within one rotation. This rotation will last less than 1 min for a 3 mm-thick, chopped spinach paste, 6 min for 10 mm-thick plaice fillets and 16-min for 20 mm-thick cod fillets. A product through-put of at least 180 kg h^{-1} is required for economical operation, and capacities up to 1000–1200 kg h^{-1} are possible. A typical operation for freezing fish fillets is shown in Fig. 7.

The product, if fluid enough to be moved by an ordinary mono-pump, is fed longitudinally through an oscillator onto the top of the freezing drum; the position of the nozzle can be adjusted for optimal surface utilisation. The stroke and the speed of the oscillating

Fig. 7. Drum freezer for fish fillets. (Courtesy of A/S Atlas, Denmark.)

movement is adjustable, and the aperture of the nozzle, made of stainless steel, can be changed according to the viscosity of the product. The movement of the oscillator is powered pneumatically at an air pressure of 7 bar (maximum). The freezing drum, which is a double-walled drum made of stainless steel, is cooled from the inside by a secondary refrigerant circulated in a helicoid; this arrangement ensures the highest possible heat transfer coefficient. The secondary refrigerant is used to avoid the high pressures associated with a primary refrigerant, for in the latter case, the drum would need an extremely thick wall, and the corresponding temperature.drop would be unacceptable. The most widely employed secondary refrigerants are either 99% ethyl alcohol (C_2H_5OH) or R11. The drum is supported by a closed, easy-to-clean, stainless steel frame, and rotates on bearings mounted at each end. At one end, the driving mechanism, consisting of a variable-gear motor is placed, and this combined with a worm gear, gives the drum an infinitely variable speed. This lay-out can be seen in Fig. 8. The device for releasing the frozen product from the drum surface is a specially profiled, stainless steel bar (Gervin-Anderson, 1977).

Fig. 8. Drum freezer drive mechanism. (Courtesy of A/S Atlas Denmark.)

Cryogenic Refrigerant Systems

Cryogenic freezing exposes the product directly to an atmosphere of liquid nitrogen (LN_2) or liquid carbon dioxide (LCO_2). These are total loss systems, and the refrigerant is discharged into the atmosphere.

Cryogenic freezers

The application of either liquid nitrogen or liquid carbon dioxide employs similar techniques, and either belt feed tunnels, spiral freezers or roll-in trolley cabinets are suitable enclosures. Figure 9 shows a roll-in cabinet modified for liquid nitrogen freezing. The cryogen is sprayed into the freezing system at a temperature of $-196°C$ in the case of liquid nitrogen, or at $-78·5°C$ liquid carbon dioxide. The liquids quickly evaporate, and the vapours are circulated by means of fans. Figure 10 shows a liquid carbon dioxide spiral freezer.

Fig. 9. Batch tray freezing cabinet using liquid nitrogen. (Courtesy of BOC plc.)

Fig. 10. General arrangement of a CO_2 spiral freezing system. (Courtesy of Distillers plc.)

The difference between the two cryogens is that carbon dioxide cannot exist in the liquid phase at atmospheric pressure, for at this pressure, it sublimates from solid to vapour at $-78.5°C$. Thus, when the liquid is discharged through the nozzles, the pressure quickly reduces to atmospheric, and about 50 per cent of the CO_2 is changed into very small particles of snow. These particles very quickly absorb heat from their surroundings and are converted into gas, and this conversion of the snow into gas will absorb $268\,\text{kJ}\,\text{kg}^{-1}$. The whole $1\,\text{kg}$ of liquid CO_2 will then have been converted to a gas at $-78.5°C$, and, if the gas is discharged from the system at $-18°C$, the total heat absorbed will be $313\,\text{kJ}$ (De Giacomi, 1972).

Liquid nitrogen systems again spray the liquid into the tunnel atmosphere, in this case at $-196°C$, and the liquid evaporates into a gas with a latent heat of $198\,\text{kJ}\,\text{kg}^{-1}$; the increase in vapour temperature will add $180\,\text{kJ}\,\text{kg}^{-1}$, giving a total heat utilisation, if discharged at $-18°C$, equal to $378\,\text{kJ}\,\text{kg}^{-1}$.

For both systems it is important that good circulation is maintained within the freezing equipment, and, to improve efficiency further, a pre-cooling section may be incorporated so that the waste vapours cool the product prior to freezing. Because of its extremely low evaporation temperature at atmospheric pressure, liquid nitrogen cannot be contained in pressure vessels, and some loss occurs during distribution and storage. However, the loss is very low for carbon dioxide because it can be contained in vessels at moderate pressure.

Although the cryogenic method is more expensive than conventional air-blast freezing, the low initial investment and simple operation makes this type of freezer economical overall, particularly as the processing plant does not require a separate installation for refrigeration.

Liquid Refrigerant Freezing Systems

The LRF (Liquid Refrigerant Freezing) process takes the product into the LRF freezer by an inlet conveyor, and then drops it into a moving stream of refrigerant at −30°C in what is designated the IQF (Individual Quick Freezing) pan. This step in the process has a three-fold purpose:

1. to separate and 'crust' freeze the individual food particles;
2. to move the product uniformly from the drop zone; and
3. to distribute the food evenly onto the freezer belt.

On transfer to the conveyor, the food passes under sprays of refrigerant at −30°C to complete the freezing, and is then carried out of the unit on a third belt. Refrigerant, which has been vaporised as a result of heat extraction from the food, is condensed by contact with a condenser located above the freeze conveyor; the refrigerant temperature in this condenser is normally −42°C. Condensed refrigerant is collected in the freezer sump and recycled to the IQF pan and spray nozzles. The overall process is shown in Fig. 11, and in Figs 12 and 13 the treatment of some individual products can be seen.

In this process, it is essential that the food is transferred from a 100 per cent air to a 100 per cent refrigerant vapour atmosphere, and vice versa, with the minimum of air/refrigerant vapour mixing. In order to achieve this aim, advantage is taken of the heavy vapour to displace the air surrounding the food as it enters the process, and conversely, the heavy refrigerant vapours are permitted to 'drain' from the food at the process exit. For most products, dissipation of residual refrigerant

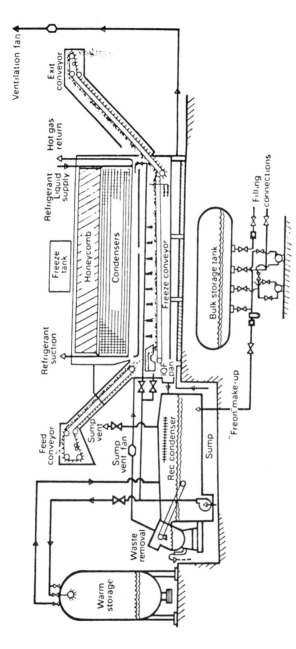

Fig. 11. Schematic arrangement of refrigerant spray freezing system. (Courtesy of E. I. Du Pont De Nemours Int. SA.)

M. F. G. Boast

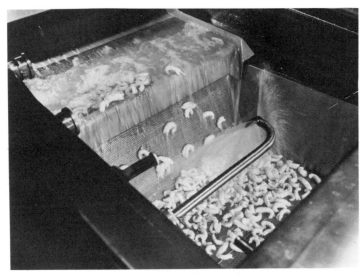

Fig. 12. Prawns being sprayed with liquid refrigerant. (Courtesy of E. I. Du Pont De Nemours Int. SA.)

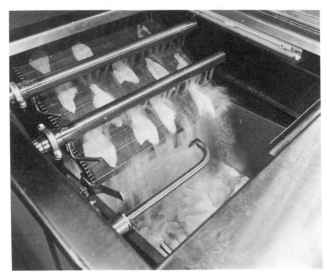

Fig. 13. Fish fillets being sprayed with liquid refrigerant. (Courtesy of E. I. Du Pont De Nemours Int. SA.)

from the food is not a problem, but with products having a large surface to mass ratio, or non-uniformity of size, a refrigerant vapour purge can be provided on the exit belt. The temperature of the vapour purge is warm enough to vaporise surface refrigerant, but not warm enough to thaw the food. The efficiency of the condensing system is also important, for during the operation of a typical freezer with wet particle products, about 250 kg of refrigerant is vaporised during the freezing of 100 kg of food. About 99 per cent of these vapours are recovered by condensation to liquid, and then re-used.

Refrigerant freezers are available in capacities ranging from 500 to 10 000 kg h^{-1}, and the usual refrigerant is dichlorodifluoromethane (R12). If dichlorodifluoromethane is used as a freezant, it must be specially purified in order to meet the specifications for acceptance as a food additive. There are two principle ways in which a person may be exposed to the refrigerant, i.e. through ingestion of very small amounts of refrigerant with foods, or by inhaling as a direct result of loss of refrigerant vapour from foods during storage or preparation for consumption. Considering ingestion, Fig. 14 is a typical curve showing the dissipation or volatilisation of refrigerant from a food in cold store. It relates to green beans, but similar curves are generated for all products frozen in refrigerant. Because R12 has such a low boiling point, it does not adhere to foods even at $-25°C$, and, as a result, most of the refrigerant is dissipated in a few days, i.e. before foods have left initial frozen storage. In this case, about 900 ppm of refrigerant were

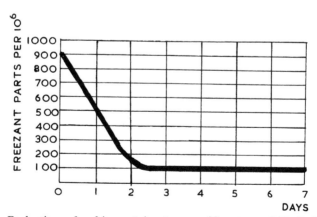

Fig. 14. Reduction of refrigerant in storage. (Courtesy of E. I. Du Pont De Nemours Int. SA.)

originally present in the green beans, but after two days in storage, only about 150 ppm remained in the frozen product. Further dissipation occurs when products are thawed or cooked for consumption, and, in the case of the green beans, cooking reduces the refrigerant residue to 20 ppm.

The time required for freezing a variety of products from room temperature to $-25°C$ ranges from only 30 s for small particulate foods (such as peas) to about 10 min for larger products. It is also claimed that there is no dehydration of the food during the freezing process, so giving higher yields than with some other systems. Adequate ventilation should be provided around the plant in order to maintain refrigerant levels in air at, or below, the threshold limit set by the appropriate authority, and under no circumstances should individuals be exposed directly to the concentrated refrigerant vapours; in particular, this applies to entering or projecting the head into a freezer containing refrigerant vapours. Body contact with liquid refrigerant must also be avoided as severe frostbite may result, and suitable goggles should be worn when there is a possibility of splashing liquid refrigerant.

Dichlorodifluoromethane will decompose at high temperatures, such as those associated with gas flames or electric heaters, and the products of decomposition are toxic and corrosive. Fortunately, these products have a very sharp, acid odour at concentrations below their toxic level, but the metallic surfaces of such heating units can be corroded by contact with the refrigerant vapours (Daly, 1973).

THE COLD CHAIN

The term 'cold chain' refers to the sequence of operations under which frozen foods are held and handled, in turn, by the producer, distributor or retailer. The intention of the system is to ensure that the temperatures maintained during the distribution of products, until they reach the consumer, are those consistent with the maintenance of high quality.

Primary cold stores are used for the long-term storage of frozen foods, and the air temperature should be between $-26°C$ and $-29°C$. Excessive temperature fluctuations, either in range or frequency, are undesirable as they may lead to some loss of consumer appeal. The intake into a primary cold store, whether from an adjacent factory or delivered from some other place, must be made with the minimum exposure of the frozen foods to outside temperature conditions. During the unloading of the vehicle, frozen foods should not be exposed to direct sunlight, wind or rain, and as much of the operation as possible (e.g. posting, strapping of pallets, etc.) should be done in the store with the packages stacked to allow a free circulation of air. It is recommended that sufficient time is allowed in the primary cold store for the product temperature to reach −23°C or colder prior to despatch. Outloading should take place with the minimum exposure of foods to outside temperature conditions, and again much of the operation (e.g. removal of pallet posts, etc.), should be done in the cold store. The temperature of the product should be checked to ensure that it is −23°C or colder, and secondary and distribution cold stores should be designed to achieve an air temperature of −23°C or less. Recording thermometers are considered essential to provide a continuous record of air temperatures, and clearly defined actions should be laid down to deal with temperature variation or refrigeration breakdown. If the air temperature rises to −12°C, it is recommended that the cold store be closed, and no further intakes or despatches should take place until the fault has been rectified and the air temperature has returned to normal. Frozen foods in store during such periods should be checked by quality control staff before being released (UKAFFP, 1978).

HOME FREEZING SYSTEMS

During the last few years, the number of homes possessing a deep-freeze has risen considerably, and now more than 50 per cent of households in the UK have facilities for storing frozen foods, and 30 per cent the capability of freezing and storing foods.

Storage Systems

A standard has been devised for the length of storage permitted for foods purchased in the frozen condition, and then quickly transferred to the freezer or frozen food section of the family refrigerator. Frozen foods should be well wrapped during transfer from the retail outlet, either in insulated bags or wrapped in a number of layers of newspaper. Partly defrosted goods should not be refrozen, but should be allowed to thaw for immediate use. The star system, using a snow flake, is used to denote the storage ability of the freezer in respect of its temperature, and the corresponding star on the purchased product will indicate the maximum length of time for which storage is recommended if quality is not to be impaired. See Table II.

TABLE II
Freezer storage ability

	Storage temperature	*Storage time for most commercially frozen foods*
Household refrigerators		
✳	Not warmer than −6°C (+21°F)	Up to 1 week
✳✳	Not warmer than −12°C (+10°F)	Up to 1 month
✳✳✳	Not warmer than −18°C (0°F)	Up to 3 months
Food freezers		
✳✳✳✳	Not warmer than −18°C (0°F)	Up to 3 months

Home freezers

Two types of home freezer are available: either with vertical or with horizontal access/loading facilities. Both types have merits, and consideration should be given to the type best suited to the application.

Chest type freezers

These cabinets are top opening and economical to operate, since even with frequent door openings, losses of the maintained temperature do not readily occur; the dense, cold air is retained. They are often provided with baskets held in suspension at a convenient working height, so giving good access to products required regularly, and long-term storage can be arranged below the baskets.

Cabinets in excess of 200 l usually have a separate freezing section and override switch, and use of the latter gives continuous operation of the refrigeration systems so depressing the temperature in this section. When home freezing with this type of freezer, the fast-freeze switch should be selected 1 h before the product is placed in the freezing compartment. In addition, this compartment should be kept empty of frozen product, since when unfrozen food is admitted, the temperature in this section will rise, and if this occurs a number of times, the quality of the stored products will be impaired. Indeed, if a chest freezer is used for regular freezing, consideration should be given to having a separate freezer just for storage. As soon as the product is frozen, the freezer should be returned to normal operation.

The volume of food placed in the freezing compartment should not exceed the quantity recommended by the manufacturer, and the items should be evenly spaced and, if possible, placed in direct contact with the freezing surface. Chest freezers are available from about 100 l up to around 600 l for commercial/domestic use, and up to a maximum of about 900 l for commercial/catering applications. Defrosting should be carried out at regular intervals so as to ensure that the walls of the cabinet are relatively free from ice, which will reduce the freezing rate

Fig. 15. Vertical domestic freezer. (Courtesy of LEC Refrigeration plc.)

of normal foods and the sub-cooling of pre-frozen products. Thin layers of ice can be scraped-off with a special scraper provided by the manufacturer, but a complete defrost and wash should be carried out every 6 to 12 months.

Upright freezers

Many of the comments applicable to chest freezers apply to vertical cabinets. However, in general, it will be found that the running costs

Fig. 16. Combined domestic refrigeration–freezer. (Courtesy of LEC Refrigeration plc.)

will be higher for the same storage volume, since when the door is opened the losses of cold air are higher; the cold air 'falls out', and is replaced by warm, moist air which increases the build-up of frost. Figures for power consumption for the same volume (around 350 l) are: chest freezers, $1 \cdot 7 \, kWh \, 24 \, h^{-1}$; and vertical freezers, $1 \cdot 9 \, kWh \, 24 \, h^{-1}$, for normal domestic family use. (See Fig. 15.)

Upright freezers may be provided with pull-out drawers, which give excellent access to stored products, and, if the drawers have closed fronts, losses of cold air are reduced. Others may only have a drop-down door or wire grid, which while less convenient, does allow an evaporator plate to form the shelf, so enabling contact freezing to take place. Freezing by placing the product in a drawer without contact with the refrigerated plate is not good practice, but again quick freeze switches are often provided, and these should be used as described for chest freezers. Door racks are not recommended, since the product can warm very quickly when access to the freezer is made. Regular checks should be made on the maintained temperature of all freezers, and routine inspection of the freezing surfaces is required to monitor the build-up of ice.

Refrigerator/freezers

These consist of a normal temperature refrigerator mounted above a vertical freezer. Normal preference is to have the refrigerator as the top section, since access is required more often to this section. A typical arrangement is shown in Fig. 16. For good temperature control and reliability, the preference is to have two independent refrigeration systems, each with independent temperature control.

REFRIGERATED TRANSPORT

The transport of frozen foods is usually in insulated and refrigerated containers or vehicles, and containers are often capable of inter-model transfer by road, rail, sea and air. The expansion of this facility has resulted in special terminals being constructed at ports so that containers can be parked, and then linked to a local electricity supply in order to maintain their contents in a frozen state. These facilities are essential for the international traffic in frozen foods.

Refrigerated trucks and containers fall into four major operational classifications:

1. Road—long-haul.
2. Distant marine.
3. Road—short-haul.
4. Local distribution to retail outlets.

Selection and Use of Refrigerated Vehicles

Vehicles should be designed and operated so that the temperature requirements of the product can be met, and a balance has to be struck between the level of insulation required to minimise heat transfer through the body, and the desired level of refrigeration. In practice, a maximum heat leakage of 0.4 W m^{-2} °C^{-1} is recommended, and vehicles should be regularly inspected to ensure that insulation and refrigeration performance are kept at the highest level of efficiency. Particular attention should be paid to door seals, punctures in the inner and outer skins and correct maintenance of the refrigerating system. Duckboards and dunnage battens should be provided to ensure free circulation of air.

Vehicle refrigeration equipment is designed to maintain, not to reduce, the temperature at which frozen foods are loaded, and any vehicle used for refrigerated transport should be equipped with some type of indicating thermometer. The instrument used most often is a remote reading, dial thermometer mounted on the exterior surface of the vehicle, preferably where the operator can see it from the cab. Often, the thermometer is combined with a control instrument which activates the fans, or the refrigeration unit, to provide control over the temperature of the vehicle interior, and there is an increasing use of temperature recorders. The sensing bulb for a remote reading recorder should be installed, for any vehicle with adequate forced air circulation, in the return air-stream in the cargo space. In a gravity circulation vehicle, or one with inadequate forced circulation, the thermometer bulb should be in a protected location about midheight in the cargo space of the vehicle. It should be remembered that a thermometer can only indicate the temperature at its sensing element, and does not assure proper temperatures throughout the cargo space; attention must also be given to keeping instruments in calibration.

Primary distribution (trunk) vehicles are for carrying frozen foods from one cold store to another, when normally only one delivery is involved, and these vehicles should be capable of maintaining frozen foods at the same temperature as those in the primary cold store, i.e. between $-26°C$ and $-29°C$. Refrigeration methods include the use of solid carbon dioxide—'dry ice'—normally regarded as a non-thermostatically controlled, total loss system, and liquid nitrogen—a thermostatically controlled, partial loss system. The latter has a heat absorption capacity of about $372 \, kJ \, kg^{-1}$, and reverts to its gaseous state at $-196°C$. Safety precautions, as recommended by the supplier, must be observed with both systems (UKAFFP, 1978). Mechanical refrigeration systems can be self-contained, and may be driven by a power unit using petrol, diesel or liquid petroleum gas.

Secondary distribution (radial) vehicles are used for the final delivery to the point of sale. Radial vehicles are engaged on multi-delivery work, and the refrigeration capacity required is dictated by the duration and frequency of door openings, rather than by heat transfer through the vehicle body. They should be designed and operated in such a way that the frozen foods can be delivered at $-15°C$ or colder. The choice of refrigerating system is governed by the need for a rapid recovery of air temperature after door openings, and for this recovery to be accomplished many times throughout the journey. The refrigeration methods listed for primary vehicles are applicable, except for 'dry ice' systems which may not achieve good temperature recovery; further details of these methods are given below.

Refrigeration Systems

Liquid nitrogen or liquid carbon dioxide spray

In this technique, the liquid refrigerant is introduced directly into the cargo space through distributing lines and nozzles. Liquid nitrogen or carbon dioxide are the principal refrigerants being used in this fashion, and their introduction can be controlled manually by hand valves, or automatically by thermostat-operated valve systems. For road operation, the refrigerant is carried in storage vessels located either inside or outside the refrigerated space, and, to eliminate the need for high pressure nitrogen tanks, the storage vessels are insulated and equipped with relief valves vented into the cargo space. Liquid carbon dioxide systems must be designed to prevent blockage of the nozzles by snow formation.

Eutectic plates for hold-over—depot charging only

Most retail service trucks use some type of hold-over plate arrangement. A hold-over plate consists of a coil for the primary refrigerant, mounted inside a thin tank of eutectic solution with a freezing temperature sufficiently low to meet the required conditions. Several of these plates are mounted on the walls or ceiling of the vehicle, or are used as shelves or compartment dividers, to provide the required refrigerating capacity while the vehicle is away from the station where the plates are refrozen. When the vehicle is garaged, flexible connections from a stationary refrigerating plant are attached to connecting devices provided on the truck; alternatively, a vehicle mounted condensing unit may be connected to an electricity supply via a trailing lead.

To provide proper cooling capacity, the plates must be mounted so that air can move freely on both sides, usually 30 to 50 mm from walls, and with top edges 150 to 200 mm from the ceiling. On ceilings, the plates should be sloped at least 1 mm in 12 across the shortest dimension, and should be at least 50 mm below the ceiling at the higher edge. Hangers must be securely fastened to the walls, preferably by fastenings incorporated in the studs, and fastenings must not penetrate through to the outside wall surface. A drip tray may be desirable under each plate, but many vehicles are defrosted by scraping away the frost without warming-up the plates. If it is necessary to protect plates from damage, guard rails should be placed at least 25 mm away from the plates. Eutectic plate systems usually circulate air by gravity, but some have fans with ducts and dampers for temperature control. The length of time that the vehicle is to be away from the depot is critical to the design and selection of the system, for sufficient time must be spent in the garage for the eutectic solution to freeze.

Eutectic plates for hold-over—vehicle-mounted condensing unit

Some trucks used for frozen food distribution are equipped with condensing units having both an auxiliary drive for operation when the vehicle is in use, and an electric motor drive for use in the garage. Installation of the plates is essentially the same as for systems charged by stationary plants, except that trucks equipped with a condensing unit which can be operated in service, can have built-in a plate eutectic capacity that is adequate for normal operation, and then operate the condensing unit for extra capacity during extended load periods. Electric motor drive units, which use standard AC supplies, are dead weight during truck operation, but make it possible to re-freeze the plates wherever electric power is available. Light-duty hold-over plates are used in some trucks which have condensing units driven by the truck engine, for these then provide a continuous source of refrigeration during periods when the truck is not operating, such as during delivery stops; an electric clutch (magnetic) controls the temperature of the storage space during long runs.

Mechanical refrigeration—independent engine or electric motor driven

Many styles of independent engine and/or electric motor driven mechanical refrigerating units are available, and these constitute the most popular form of mechanism for obtaining low temperatures.

Fig. 17. A hydraulic–electric drive system. (Courtesy of Hubbard Engineering.)

One example of this type is the one-piece, plug-type design, which is a self-contained unit mounted in an opening provided in the front wall of the vehicle. (See Fig. 17.) The condensing unit is on the outside, and the evaporator section on the inside, and the two sections are separated by an insulated 'plug' attached to the vehicle wall; this plug also supports the various parts of the refrigerating unit. Some of the

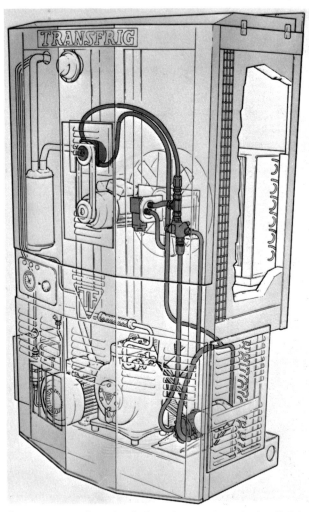

Fig. 18. Arrangement of mechanical equipment for a hydraulic/electrical unit for trailers. (Courtesy of Transfrig Ltd.)

units are tall and slender so as to mount between the tractor and trailer, others are short and deep and extend over the tractor roof. Some units have compressors that are belt-driven, or direct-driven, by an auxiliary engine; and these engines may operate on fuel from the vehicle tank, or from a separate fuel supply. Many units are equipped, in addition, with an electric motor for standby operation, and are thermostatically controlled, starting and stopping or reducing speed as the refrigeration need requires. Condenser and evaporator fans are driven by the unit engine, and tight-fitting dampers close-off the evaporator air-flow to permit defrosting, by the hot gas method, without unduly warming the vehicle interior; most units defrost automatically. The one-piece construction facilitates unit replacement in the event of breakdown, or removal for maintenance.

Mechanical refrigeration—power take-off from vehicle engine or transmission

For all practical purposes, this type of refrigerating system is limited to larger vehicles, and several means of power take-off are available. (See Fig. 18 for general arrangement, and Fig. 19 for a mounted unit.)

One system uses an alternator, belt-driven from the engine

Fig. 19. Trailer mounted hydraulic/electric drive system. (Courtesy of Transfrig Ltd.)

crankshaft, to provide a regulated alternating current voltage which is rectified to drive direct current motors for the compressor and fans in the body. Temperature control is accomplished by cycling the compressor, but refrigeration is produced only when the truck engine is in operation. Another system has a condenser–evaporator assembly mounted in the nose of the truck, with the compressor driven directly by a power take-off at the transmission. When the vehicle is garaged, an electric motor is used to drive the compressor, and the current is rectified to operate the direct current condenser and evaporator fan motors. A number of different types of hydraulic clutch have been used, with primary power taken from the truck transmission or drive shaft, and as with other systems using power take-off from the truck engine, the use of eutectic plates provides continuous refrigeration when the truck engine is not operating. Since there are many combinations of features and styles of power take-off equipment available, prospective users are advised to study their particular needs carefully in order to take advantage of the wide selection (ASHRAE, 1982).

Vehicle Loading

Frozen foods are most vulnerable to temperature rise during the loading and unloading of distribution vehicles, and the following methods are recommended to keep temperature changes to a minimum:

—Pre-cooling of vehicles before loading.
—The use of loading/unloading ports, with the vehicle in direct contact with the entrance of the cold store.
—Where ports are not available, covered bays should be used, and the vehicles should be screened from direct sunlight, wind and rain.
—Minimise loading/unloading times by means of pallets and mechanical handling equipment.
—Sorting and pre-assembly of loads should be carried out in the cold store.
—Unloading should be as rapid as possible.

REFERENCES

ASHRAE (1982). *Application Handbook,* American Society of Heating, Refrigeration and Air Conditioning Engineers, Atlanta, USA.

Boast, M. F. G. (1982). *J. Society of Environmental Engineers*, **2**, 57.
Cameron, H. C. (1904). *Proc. Institute of Refrigeration*, **2**, 3.
Daly, J. J. (1973). *Food Processing Industry*, **42**, 10.
De Giacomi, R. (1972). *Food Processing Industry*, **41**, 3.
Forbes-Pearson, S. (1983). Symposium of the Institute of Mechanical Engineers/Institute of Refrigeration, 'Refrigerated storage and handling of frozen foods'. March, 1983.
Gervin-Anderson, A. (1977). *Proc. Institute of Refrigeration*, **74**, 2.
UKAFFP (1978). *Code of Practice*, UK Association of Frozen Food Producers, London.

Chapter 2

Effects of Freezing/Thawing on Foods

L. Boegh-Soerensen and M. Jul

The Royal Veterinary and Agricultural University, Copenhagen, Denmark

INTRODUCTION

Freezing preservation includes three steps: a freezing process, storage at freezer temperatures, and a thawing process, and this chapter deals with the changes—physical, chemical and nutritional—which occur during freezing preservation.

Freezing preservation is a very gentle preservation method. Chilling, properly carried out, may have less effect on quality, and cause less physical and chemical change than freezing, but when it comes to loss in nutritive value, chilling may result in larger losses than freezing, especially of the labile vitamins, e.g. ascorbic acid and thiamin. The other preservation methods (drying, curing, smoking, chemical preservation, fermentation and heat processing) result in greater changes in the foodstuffs. Most of the changes take place during freezer storage. Freezing and immediate thawing has, for a number of food products, so little influence that it is impossible to distinguish between the unfrozen product and the frozen and thawed product on the usual quality attributes, no matter what method of analysis (sensorial or chemical) is used.

Generally, in this chapter, we have used the expressions 'colder temperatures' and 'warmer temperatures' and although this is not scientifically correct, we have used them to make it clear whether we are talking about a decrease or an increase below freezing temperature.

41

BASIC ASPECTS OF FREEZING AND THAWING

Formation of Ice

It is characteristic of frozen foods that a high proportion of the water content is found as ice, and it is well-known that the quicker the cooling process is, the smaller the ice crystals will be. During freezing of foodstuffs, ice crystals begin to form in the liquid between the cells, and the main reason is presumably the higher freezing point of this extracellular liquid compared with the intracellular liquid (Love, 1968). During slow freezing, ice crystals grow between the cells (fibres) making the extracellular liquid more concentrated. The cells will lose water by osmosis, and this leads to an extensive dehydration of the cells and the cells contract. The result will be relatively few, large ice crystals between shrunken cells as shown in Fig. 1(b). During rapid freezing, heat is removed so quickly that there is little time for dehydration of the cells, and ice crystals will be formed also in the cells. The result could be as shown in Fig. 1(c), a number of small crystals, between and inside the cells, and an appearance of the frozen tissue that is very close to the appearance of unfrozen tissue. The

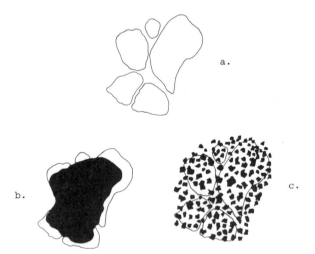

Fig. 1. Diagram of freezing of foodstuffs. (a) Unfrozen tissue; (b) the same tissue after slow freezing, with one ice crystal between dehydrated, shrunken cells; (c) the same tissue after very rapid freezing, with small ice crystals between and inside cells.

distribution and size of ice crystals in frozen foods has been the subject of many investigations; see among others Plank *et al.* (1916), Luyet (1968). It is easy to imagine that a slowly frozen product (Fig. 1(b)) will lose a certain amount of water (drip) during thawing, simply because the water may be unable to return to its original position. It is equally easy to conclude that no such problems arise in a rapidly frozen product. Thus, the widespread view of quick freezing leading to small ice crystals and superior quality of the frozen foodstuff was formed.

However, although most experimental data show little or no influence of freezing rate on food quality, this view is still popular, and is even incorporated in legislation in some countries.

Amount of Water Frozen Out

From a physical point of view, foods may be considered as dilute aqueous solutions, with a freezing point below 0°C. The freezing point depression is $1·86°C \, mol^{-1} \, litre^{-1}$ which means that the freezing point

TABLE I
Water content and approximate initial freezing point of some foods. Adapted from Spiess (1977), Boegh-Soerensen *et al.* (1978), and Polley *et al.* (1980)

Product	Water (%)	Initial freezing point (°C)
Meat, lean	74	−1·5
Meat, cured, 3% NaCl	73	−4
Fish, lean	80	−1·1
Fish, fat	65	−0·8
Poultry	74	−1·5
Egg yolk	50	−0·7
Egg white	87	−0·5
Milk	88	−0·5
Lettuce	94	−0·4
Tomatoes	94	−0·9
Cauliflower	92	−1·1
Peas	76	−1·1
Potatoes	78	−1·7
Strawberries	89–90	−1·2
Apples	84–86	−2
Cherries	86	−4
White bread	35	−5

depends on the concentration of dissolved molecules in the water phase, and not on the water content in the foodstuff. Table I shows the initial freezing point and the water content for some foods. The freezing point of water is clearly defined, as all water freezes to ice at 0°C, but for foodstuffs the denomination 'initial freezing point' has to be used, since the more the product is cooled below initial freezing, the more water will freeze out, making the remaining water phase more and more concentrated. This will lower the freezing point, requiring a progressively lower temperature to freeze the residual water. Even at −180°C, perhaps even lower, there seems to be a small amount of unfrozen water (Love, 1968). For some frozen foods, the percentage of

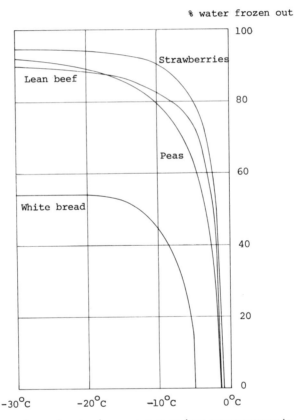

Fig. 2. Percentage of water frozen out at various temperatures in four foods. Adapted from IRR (1972).

water frozen out at different temperatures is shown in Fig. 2. As discussed later, the increase in concentration with increasingly cold temperatures may have some undesirable effects on food quality.

The Freezing Process

In food technology, the freezing process is a process where the temperature of the food product is lowered from the initial temperature to a temperature below the initial freezing point of the food product; it involves several stages:

(a) Prefreezing stage.
(b) Supercooling (undercooling).
(c) Freezing stage.
(d) Subfreezing stage.

These four stages can be seen in Fig. 3, showing a freezing curve for distilled water.

(a) *Prefreezing stage*, from the beginning of the freezing process until the temperature of the food product has been lowered to the freezing point.

(b) *Supercooling* or undercooling means that the product temperature drops below the freezing point without ice being formed. Distilled

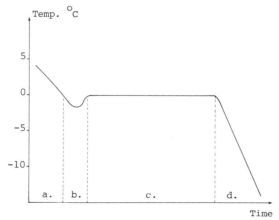

Fig. 3. Temperature during freezing of pure water.

water may be undercooled to −20°C or even lower. In commercial freezing of foods, undercooling is seldom seen, although it is believed to occur in all biological materials. Andersen *et al.* (1965) mention that eggs can be undercooled to about −3°C without any ice formation taking place. However, if the eggs are stirred mechanically, e.g. due to transport, ice formation will occur rapidly.

(c) *Freezing stage*, where water becomes ice. In Fig. 3, the freezing stage is shown as a plateau on the freezing curve, meaning that, in distilled water, the temperature remains constant at 0°C until all water has passed from the liquid to the crystalline state, i.e. to ice crystals. The freezing curve for food products differs considerably from the freezing curve for water (see Fig. 4).

(d) *Subfreezing state*, where the product temperature is lowered to the end temperature, which should be the intended storage temperature. This part of the process is relatively rapid, as ice has a lower specific heat than water, and practically no removal of latent heat is involved.

Freezing Curves in Food Products

In freezing food products, there will normally be a pronounced difference between the freezing curve of the surface and that of the centre as shown in Fig. 4. The surface temperature drops continually, with no plateau. In the thermal centre, there will be a plateau where water changes phase to ice, but the plateau is not horizontal, since the freezing point decreases as the concentration of the remaining solution increases. Figure 4 refers to blast freezing of a hindquarter of beef with a diameter of 26 cm, and, if the hindquarter, after blast freezing for 16 h, is placed at a temperature of −20°C, in still air, the surface temperature would be −21°C and the centre temperature −19°C after 4 h of equilibration.

The picture will vary in response to a number of factors, including the freezing process, the size of the food product and the packaging.

Temperature

After storage at relatively constant temperature for some time, the temperature in a product will have equilibrated and will thus be rather uniform and, therefore, easier to measure. As a matter of fact, in large

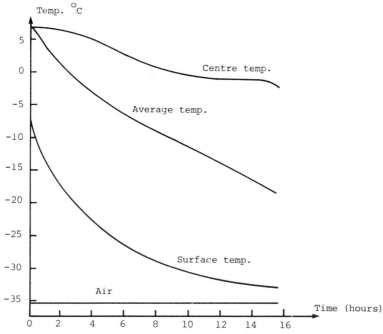

Fig. 4. Temperatures during freezing of a big piece of meat. After Fleming (1974).

storage rooms with a good air circulation and air distribution, product temperatures and air temperatures do not differ very much after some days. During freezing and thawing, there will be great differences between the temperature of the surrounding medium, and the surface and centre temperatures of the product (see Figs 4 and 8). The centre temperature is normally defined as the temperature in the thermal centre of the food product. Therefore, it should be comparatively simple to measure by placing a thermocouple in the centre. In practice, it is not always that simple, e.g. in chickens or peas. By 'product temperature' the equilibrium (equalisation, equilibration) temperature or average temperature should be understood, i.e. the temperature that the product will attain after thermal stabilisation has been achieved under adiabatic conditions (IIR, 1972). Adiabatic conditions are those under which no heat is either added or abstracted, e.g. the conditions obtained by placing the food product in an insulated box.

The equilibrium (average) temperature normally differs considerably from the centre temperature during freezing (see Fig. 4).

The freezing process should be regarded as completed when the equilibrium temperature has reached the intended storage temperature, because the necessary amount of heat has been removed from the product. If, as in Fig. 4, the food product is going to be placed in a freezer storage room at $-20°C$, the freezing process could be stopped after 16 h although the centre temperature is as warm as $-3°C$; there is absolutely no need to continue the freezing until the centre has reached $-20°C$. On the contrary, continued cooling of the product may result in unnecessary low temperatures in the surface layers. This may be harmful to product quality, and result in energy waste and inefficient use of freezer capacity. In practice, the freezing process is often regarded as completed when the centre temperature has dropped to $-10°C$. Figure 4 shows that, in some cases, even warmer centre temperatures could be sufficient. A special technique is required to measure average temperature, and a specific time must elapse— depending on the size of the food product, e.g. about 3 h for chicken—before the temperature can be read. As energy consumption is much higher during freezing than during freezer storage, it could pay to clarify the relationship between the centre temperature and the equilibrium temperature for each product to be frozen. In this way, one may, for different conditions and products, be able to determine the optimum freezing time, and thus the most efficient method of using the freezing installation.

Freezing Time

A large range of freezing equipment is available, some rather universal, some tailor-made for the freezing of specific products. The choice of freezing equipment—freezing method—depends on several factors, including cost, weight loss during freezing, flexibility, reliability, and the desired freezing time. A rapid freezing method, e.g. an in-line freezer, may result in better product quality and in economical advantages, due to higher through-put. A rapid freezing method may also be preferred by the microbiologist since at temperatures colder than about $-10°C$, no microbiological activity takes place until the product is thawed.

The freezing time can be defined in several ways.

Definition No. 1 (IIR, 1972): The effective freezing time is the time required to lower the temperature of the product from its initial temperature to a given temperature at the thermal centre'. The temperature at the thermal centre is often set at −10°C.

Definition No. 2: In IIR's definition of freezing rate, freezing time is defined as the time from the surface reaching 0°C to the thermal centre reaching −10°C.

In air-blast freezing of cartoned products stacked on pallets, some time will elapse from the start of the freezing until the surface temperature of the products in the centre has dropped to 0°C. This period may at times be long enough for micro-organisms to grow, which could influence the storage life. Thus, the freezing of rather warm products, in less than perfect microbiological condition, can lead to the product being unacceptable even before freezing actually sets in (Löndahl and Nilsson, 1978). In such cases, definition No. 1 of freezing time is a more realistic concept than definition No. 2. In the frozen food industry, the (effective) freezing time varies from a few minutes, e.g. in fluidised-bed freezing of peas, to one or more days, e.g. in air-blast freezing of cartoned meat.

Freezing Rate

It is not very meaningful to compare freezing times for products of vastly different size, e.g. beef quarters and peas, and hence the concept of freezing rate has been introduced. Freezing rate can be expressed in several ways. Temperature change per time unit, e.g. $°C\,s^{-1}$, is sometimes used, but, as can be seen from Fig. 4, the temperature change (drop) will vary considerably from the surface to the centre, thus making this approach less useful for characterising certain freezing processes. Freezing rate is normally expressed, therefore, as the average velocity at which the ice front advances from the surface to the thermal centre, and, for practical purposes, an average freezing rate can be defined as the ratio between the minimum distance from the surface to the thermal centre and the freezing time. Where depth is measured in centimetres and freezing time in hours, the freezing rate will be expressed in $cm\,h^{-1}$; for freezing time, definition No. 2 is used.

Freezing methods may be characterised by the freezing rate, as

TABLE II
Freezing rate of different freezing methods.
After Boegh-Soerensen *et al.* (1978)

Freezing method	Freezing rate $(cm\,h^{-1})$
Ultra rapid freezing	over 10
Rapid freezing	1–10
Normal freezing	0·3–1
Slow freezing	0·1–0·3
Very slow freezing	less than 0·1

shown in Table II, and the usual categories are:

1. *Ultra rapid freezing* may be achieved by freezing small-sized products in liquid nitrogen, liquid fluorocarbon or carbon dioxide.
2. *Rapid freezing* can take place in fluidised-bed freezers and in plate freezers.
3. *Normal freezing* is found in most air-blast freezers.
4. *Slow freezing* is found during air-blast freezing of cartoned meat.
5. *Very slow freezing* can result when the air speed around the products in a blast freezer is too low, or when the air temperature is too warm, or when a product is frozen in units which are too large.

Properties of Frozen and Unfrozen Foods

The thermophysical properties of foods depend on temperature, and frozen and unfrozen products have quite different properties.

The thermal conductivity of water is about $0\cdot5\,W\,m^{-1}\,{}^{\circ}C^{-1}$ and that of ice, depending on the temperature, is about $2\cdot4\,W\,m^{-1}\,{}^{\circ}C^{-1}$ (see Fig. 5, which also shows the thermal conductivity for lean beef and fat).

The specific heat for water is $4\cdot19\,kJ\,kg^{-1}\,{}^{\circ}C^{-1}$ (1 kcal 1 $kg^{-1}\,{}^{\circ}C^{-1}$), i.e. to change the temperature of 1 kg water 1°C, it is necessary to remove or add 4·19 kJ. For ice, it is about $1\cdot8\,kJ\,kg^{-1}\,{}^{\circ}C^{-1}$ (0·45 kcal $kg^{-1}\,{}^{\circ}C^{-1}$). As most food products contain large amounts of water, the specific heat for food products will be rather close to that of water or ice, i.e. 65–80% of that of water or ice. Figure 6 shows the specific heat for water, ice and lean beef.

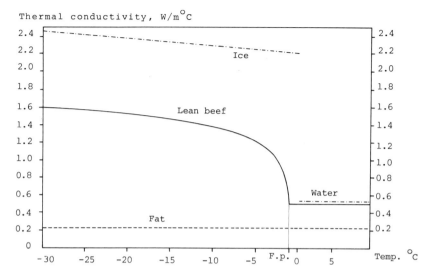

Fig. 5. Variation of thermal conductivity with temperature for water, beef and fat. Adapted from data compiled by Morley (1972) and Mellor (1978).

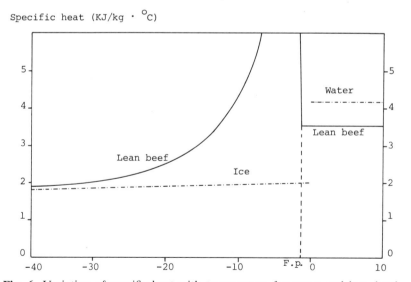

Fig. 6. Variation of specific heat with temperature for water and lean beef. Adapted from Morley (1972) and Mellor (1978).

TABLE III
Approximate specific heat at temperatures above freezing
point of components in foodstuffs. After Kostaropoulos
(1979)

Component	Specific heat $(kJ\ kg^{-1}\,°C^{-1})$
Water	4·2
Protein	1·9
Fat	1·9
Carbohydrate	1·2
Metals	0·8

The specific heat, c_p, can be estimated by means of the formula:

$$c_p = c_i m_i \tag{1}$$

where c_i is the specific heat $(kJ\ kg^{-1}\,°C^{-1})$ of each component of the food product and m_i is the fraction $(g\ kg^{-1})$ of each component. The approximate specific heat of the main food components can be seen in Table III. The specific heat of foodstuffs containing more than 25 per cent water is $2·7–3·9\ kJ\ kg^{-1}\,°C^{-1}$, above the freezing point, and $1·5–2·3\ kJ\ kg^{-1}\,°C^{-1}$, below the freezing point, in both cases dependent on the water content.

Mellor and Seppings (1976) regard the foodstuff as a mixture of water and solids, and thus calculate the specific heat by

$$c_p = 4·18m_w + 1·38(1 - m_w) \tag{2}$$

where m_w is the fraction of water; $1·38\ kJ\ kg^{-1}\,°C^{-1}$ is the specific heat of the solid content of most foodstuffs. Such calculations lead to values within 10 per cent of measured values.

Such formulas may be used down to the freezing point, but, with regard to frozen foods, the calculation is not straightforward because of the fact that there is always a proportion of unfrozen water. Calculations of the necessary heat to be removed during freezing, or added during thawing, should take into account the specific heat above, at, and below the freezing points. However, the specific heat of foods is comparatively constant at temperatures warmer than the freezing point of the foodstuff or colder than −20°C, but during the freezing process, the specific heat is not constant. The reason is the amount of energy necessary to freeze water (or to melt ice), i.e. the so-called latent heat. It is difficult to estimate the latent heat as ice formation takes place over a temperature span of at least 18°C.

TABLE IV

Specific heat of unfrozen foods and enthalpy of foods at different temperatures. Adapted from Riedel (1950), Riedel (1959), Riedel (1964), Morley (1972), Riedel (1977) and Riedel (1978)

Food product	Specific heat[a] $(kJ\,kg^{-1}\,°C^{-1})$	Enthalpy $(kJ\,kg^{-1}; 0\,kJ\,kg^{-1}$ at $-40°C)$					
		$-30°C$	$-20°C$	$-10°C$	$0°C$	$10°C$	$20°C$
Beef, 75% H_2O	3·52	18·2	40·6	70·5	298·1	332·7	368·6
Pork, 2·5% fat	3·51	19·0	41·3	74·7	294·0	329·1	364·2
Pork, 25% fat	3·69	18·1	38·9	68·6	250·3	289·4	324·2
Pork, 48% fat	3·85	17·9	38·7	67·9	212·3	254·6	289·2
Cod, 80% H_2O	3·68	20·5	41·8	74·1	322·8	361·1	389·6
Herring, 20% fat	3·35	20·1	41·3	73·2	278·4	314·4	348·8
Lard, 95% fat	—	16·6	41·3	66·6	100·1	147·4	180·4
Egg white, 87% H_2O	3·62	19·7	39·8	67·0	321·5	357·5	393·9
White bread, 35% H_2O	2·48	18·0	36·0	65·7	126·8	151·8	176·2
Carrots, 88% H_2O	3·84	18·9	44·3	81·2	358·3	396·3	434·9
Peas, 76% H_2O	3·54	17·8	43·4	87·0	312·4	347·6	383·5
Strawberries, 89% H_2O	3·92	19·2	42·1	81·1	367·5	405·4	442·7
Raspberries, 83% H_2O	3·73	20·2	45·8	83·6	344·3	380·7	418·5

[a] Specific heat is an average value in the temperature range 0°C to 30°C.

Enthalpy. The above mentioned difficulties have led to the use of the total heat content (enthalpy). In Table IV, the enthalpy is given as a function of temperature, the enthalpy arbitrarily given the value 0 at −40°C.

Volume. Ice has lower density than water, 0·92 kg dm^{-3} versus 1 kg dm^{-3}. The volume increase accompanying the conversion of water into ice is about 9 per cent, and the volume increase of foods as a result of ice formation is about 6 per cent. For some fruits and vegetables with a large amount of air in the intercellular spaces, the volume increase may be about 3 per cent. This volume increase has to be taken into account, for example when deciding on the packaging method of the unfrozen product.

Internal pressure. Due to the volume change during freezing, stresses can be built up, large enough to distort the shape of the product and to crack the surface. This has been found especially during very rapid freezing, i.e. in liquid nitrogen (Lorentzen, 1964); the author measured internal pressures of about 12 bar. Miles and Morley (1977) measured higher internal stresses in meat during freezing at −50°C than at

$-10°C$, and found a maximum stress of 60 bar. Such internal pressures often result in cracks of the surface.

Thawing

For most foods, freezing and freezer storage normally results in satisfactory quality when good manufacturing practice is followed, but if a product has to be thawed prior to processing or to cooking, a number of problems may arise, because thawing is more difficult to control than freezing.

The thawing process is divided into three parts (see Fig. 7):

(a) Heating the solid to its thawing plateau (tempering).
(b) Thawing.
(c) Heating the food above its thawing point.

The overall thawing time of a product is the time elapsed from its initial frozen temperature to the point where no ice remains in the product. During the thawing process, heat must be supplied; the latent heat for melting of ice is about $334\,kJ\,kg^{-1}$. The necessary amount of energy for a given product can be found in enthalpy tables, see Table IV.

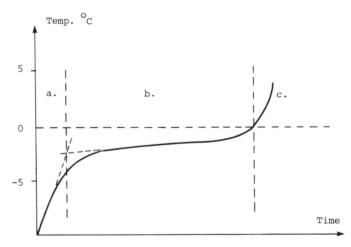

Fig. 7. Typical thawing curve for a foodstuff. After Bailey (1981).

Thawing Methods

Thawing methods can be divided into two groups: Surface heating methods, where heat, supplied to the surface by means of air, water, or condensing steam (vacuum-thawing), is conducted into the food from the surface; and internal heating methods, where heat is generated within the product, by means of electrical resistance, dielectric or microwave heating.

Surface heating methods. At first glance, air thawing in circulating 'warm' air, should be as rapid as freezing in circulating cold air. However, as shown in Fig. 8, the temperature curves are characteristically different during thawing (a) and freezing (b), although the heat transfer coefficient and the temperature difference are almost identical (Jason, 1974). The main reason is that the thermal conductivity is lower for water than for ice, so that, during thawing, an increasingly thick layer of unfrozen material will decrease the conductivity of heat from the surface to the centre, whereas during freezing, an increasingly thick layer of ice facilitates the transfer of heat from the centre to the surface. In addition, whereas the temperature of the freezing medium can be chosen as cold as practicable, the temperature of the thawing medium must be controlled so that the surface temperature does not become too high; otherwise problems can arise with the colour and appearance of some products, and with the microbiological condition

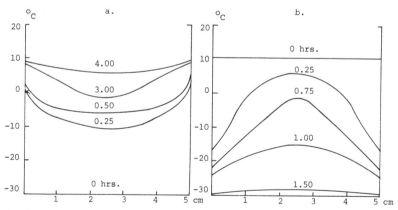

Fig. 8. Temperature profiles in a 50-mm-thick slab of fish during thawing (a) and freezing (b) at various times after commencement of the operation. After Jason (1974).

of most. In air thawing, a programmed air temperature is sometimes used. The starting air temperature may be 30–40°C, but when the product surface reaches a predetermined level, the air temperature is reduced to 5–10°C, depending on the food. When thawing unwrapped products, the relative humidity should be controlled also, in order that the water activity of the surface is kept sufficiently low to control microbiological growth, and yet high enough to minimise water evaporation, i.e. shrinkage.

Internal heating methods. Microwave thawing has been the subject of considerable interest during recent years, for thawing times can be reduced to minutes as against hours or days for surface heating methods. However, the electrical methods demand a rather more homogeneous foodstuff than is found in practice, and furthermore, the rate of heating increases as the product thaws, which makes uneven heating a great problem. Today, mainly microwaves are used, and only for tempering, or for thawing and heating frozen foods at the institutional or home level.

FREEZING PRESERVATION—INFLUENCE ON FOOD QUALITY

Few frozen products are eaten in the frozen state (ice cream, sherbets, etc.), and hence, it is not only the influence of the freezing process on food quality that is of interest, but the combined influence of freezing and thawing. Some products which are usually eaten in the uncooked condition, such as leafy vegetables like lettuce, tomatoes, and several fruits, are not suited for freezing preservation, because the freezing process leads to a breakdown of cell walls and often to other freezing injuries. However, many of these changes are quite similar to changes caused by heating, so that products intended for cooking later, e.g. spinach and pie filling, can well be preserved by freezing. For other products, e.g. meat and fish, appropriate freezing and thawing with no intervening frozen storage period leaves the product practically unaltered; thus, frozen/thawed beef may be used for making tartar, and frozen salmon for a 'gravad' (very lightly cured) product.

In general, for products that will be cooked, freezing and thawing—with no or a very short period of frozen storage—has little effect on quality, for most quality deterioration takes place during storage. The rate of quality-changing processes normally, but not always, decreases with colder storage temperatures. The nature of the most important changes caused by freezing preservation are described below.

Concentration of the Water Phase

During the freezing process, water is converted into ice crystals with a high degree of purity, leading to a concentrated solution of salts, minerals, and other substances; the extent of the concentration depends on the product, the end temperature, and, to a certain extent, the freezing rate. This increased concentration of solutes often causes a change in pH—usually towards the acid side—which can influence product quality (van den Berg, 1968).

Freezing may cause physico-chemical changes such as loss of water binding capacity, resulting in drip; protein changes leading to toughening or dryness; and loss of turgor in fresh fruits and vegetables. Many physico-chemical changes increase with increasing solute concentration in the water phase, but may, at the same time, decrease with colder temperatures. However, with colder temperatures more water will freeze out, so that, during thawing, more water will have to return to its original place. This observation could lead to a theory that very low end-temperatures during freezing should be avoided. In practice, this appears to be a problem when freezing in liquid nitrogen or carbon dioxide, i.e. at very cold temperatures.

In general, the temperature range which causes most irreversible changes is from about −1°C (the initial freezing point) to about −5°C; therefore, foods should pass this range reasonably speedily during freezing as well as during thawing. For the same reason, temperatures warmer than −10°C are normally avoided in freezer storage, although some cured meat products do not follow this pattern; on the contrary, they may exhibit a longer shelf-life at −10°C than at −24°C.

Drip Loss

Thawing is often accompanied by the release of a fluid from the frozen product. This fluid is known as drip. The amount of drip depends on the type of product and the surface area, but also on freezing method, storage conditions, and thawing method. Drip may amount to 3–5 per cent or even more, and it is natural to ascribe the somewhat dry texture of some frozen foods to the amount of drip. However, as

TABLE V

Freezing time, weight loss during freezing, and drip loss for four different
foodstuffs frozen by four different methods. After Åström and Löndahl (1969)

Product	Freezing method	Freezing time (min)	Weight loss during freezing (%)	Drip loss (%)
Strawberries	Fluidised-bed	14	1·5	8·0
	Freon	3	0·0	5·5
	Nitrogen	5	1·4	11·0
	Air-blast	900	2·7	21·0
Mushrooms	Fluidised-bed	22	1·2	30·6
(not blanched)	Freon	3	0·0	33·2
	Nitrogen	5	1·9	30·2
	Air-blast	180	2·5	30·0
Plaice	Spiral freezer	18	0·8	0·2
	Freon	4	0·1	0·3
	Nitrogen	8	1·4	0·6
	Air-blast	180	1·5	0·5
Beef fillet	Spiral freezer	35	0·7	0·2
	Freon	4	0·1	0·2
	Nitrogen	8	1·4	0·03
	Air-blast	180	1·3	0·5

described later for the different product groups, the picture is not very
clear. Some experimental results showing the influence of different
freezing methods on drip loss for four different products are given in
Table V. Fruits and vegetables may exhibit a very high drip loss upon
thawing, unless frozen very quickly and stored at very cold temperatures; meat and fish seem somewhat less sensitive. The quality of the
thawed products was tested organoleptically; air-blast-frozen products
had slightly poorer eating quality than products frozen by one of the
three 'advanced' methods.

Appearance

The freezing method often has limited or no influence on the eating
quality of the thawed (cooked) product. However, the quicker the
freezing is, the smaller the ice crystals will be, which means that,
because of optical phenomena, a quick frozen product will have a light
appearance and a slow-frozen product will be dark. After thawing,

there will be no difference in appearance between a slow- and a quick-frozen product.

A light appearance is regarded as an advantage for some products; chicken and turkey broilers must have a whitish appearance according to consumers in some countries. This is sometimes obtained by carrying out the first part of the freezing in a liquid, i.e. crust freezing, and completing freezing in a normal air-blast freezer or even a freezer storage room. The latter procedure could involve microbial hazards, if the crust freezing has not lowered the centre temperature sufficiently.

For other frozen products, e.g. red meat, most consumers would regard a whitish appearance of the frozen product as objectionable. Changes in appearance during freezer storage are a problem for many frozen foods.

Physical Changes

Physical aspects, e.g. ice formation and volume/pressure changes, and changes in such properties as specific heat and thermal conductivity have been dealt with earlier.

Desiccation

Desiccation can occur during freezing and thawing as well as during freezer storage.

Unpacked products. It is inevitable that water evaporates from unpackaged frozen foods. During the freezing process, the moisture losses vary from 0·5 to 2 per cent or more, depending on the freezing method and the product (see Table V). Fast freezing lowers the surface temperature of the product quickly to a temperature where evaporation or sublimation is nearly negligible. In air-blast freezing the air temperature should be $-30°C$ or colder, and the air velocity as high as practical to ensure a low evaporation loss. However, it should be noted that for some unpackaged products, an air velocity above 1 m s^{-1} may result in superficial colour changes of the frozen product (Clemmensen and Zeuthen, 1974).

During storage of unpackaged products, a weight loss is difficult to avoid. It decreases with colder temperatures, as colder air can contain less water vapour, so that at $-28°C$, the weight loss of unpackaged products may be half the weight loss at $-18°C$. In freezer storage rooms, air velocity should be as low as possible, taking into account that the temperature must be kept uniform, for this will also reduce evaporative loss.

During air thawing, moisture losses depend on temperature, velocity and relative humidity of the air. The velocity of the air around the products should not exceed $3\,m\,s^{-1}$, and when thawing unpackaged foods, the relative humidity should be as close to 100 per cent as microbiological conditions permit.

Packaged products. In most cases, the packaging will be a plastic pouch with a rather low water vapour permeability. During freezing, packaging in a material with a low water vapour permeability does not guarantee a low moisture loss from the product. If the packaging does not fit tightly around the product, water will evaporate from the product and deposit on the inside of the package as frost, so-called in-package desiccation. Only a tight fitting packaging material with low (or very low) water vapour permeability ensures that there will be no moisture loss. In freezer storage, the problem is more serious since the time involved is much longer. The use of packaging materials with high water vapour permeability results in a moisture loss which depends on storage temperatures and the temperature fluctuations which occur during the product's distribution.

In-package frost formation is due to temperature differences between the product surface and the packaging material. When outside temperatures are colder than that of the product, ice on or in the latter will sublime and condense on the inside of the packaging. As outside temperatures become warmer, the process is reversed, but the water vapour will not condense and return to the cellular structure from which it evaporated (sublimed). This inside frost formation can amount to several per cent of the product weight, and is a problem especially for products in transparent plastic pouches in display cabinets, because it results in a very untidy appearance. It should be noted that product temperatures in open display cabinets are rather warm and fluctuate over a considerable temperature range (Boegh-Soerensen and Bramsnaes, 1977).

Freezer burn. Desiccation leads to undesirable moisture loss, but excessive desiccation may have more serious implications. After some desiccation, small white spots are formed on the surface of, for example, chickens, and the observing consumer might think it is a growth of mould; the appearance of the thawed product is unaffected. Continued desiccation, however, results in the removal of ice from the surface layers, so allowing free access of oxygen, a phenomenon which can initiate and accelerate oxidative changes. Eventually, this leads to irreversible deterioration of taste, texture, and appearance, i.e. freezer burn, which is often seen as grey or brown discoloration of the surface. Freezer burn is especially a problem in animal foods (meat, poultry, fish).

Recrystallisation

During freezer storage, the size (and shape) of the ice crystals may change. This happens especially at rather warm temperatures, i.e. warmer than about $-18°C$, but all temperature fluctuations will inevitably cause the growth of large ice crystals at the expense of small ones (Luyet, 1968). This means that after some months in freezer storage, products frozen rapidly with a large number of small ice crystals formed, see Fig. 1(c), will now contain larger ice crystals. The influence of recrystallisation on product quality is not very well-known, but is presumably rather little.

Chemical Reactions During Freezer Storage

The freezing process results in few chemical changes, unless it is very slow freezing, and the same goes for the thawing process. However, during freezer storage, several chemical reactions may take place, resulting in changes in texture, appearance, flavour, and nutritional value; these reactions may be of a physico-chemical, biochemical (enzymatic) or chemical nature.

Texture

Optimal texture varies from food to food. In fruits and vegetables, it is the proper turgor (internal pressure) and/or crispness. In meat, poultry, and fish, it is the proper degree of tenderness. In sauces, it is uniformity; separation of the water phase is a serious texture defect.

Proteins. Protein denaturation is difficult to define exactly, because it really refers to changes in the properties of a protein. Most proteins are stable to freezing, although some enzymes are damaged by freezing. Oxygen (oxidation) presumably has some influence on the changes in proteins since vacuum packaging in most cases reduces the texture changes during freezer storage. In fish, freezing often results in a certain denaturation of protein (especially myosin), leading to toughness and dryness; the increased concentration of salt (and other electrolytes) is presumably the main explanation. The myosin of beef and poultry is much more stable. Thawed egg yolk is gummy and partially coagulated, and this change occurs at temperatures below $-5\cdot5°C$ due to lipoprotein aggregation. The effect is minimised by adding salt or sugar before freezing. The freezing of egg white has little influence on its quality, but the freezing of cooked egg white presents a serious problem. The white becomes tough, rubbery, and watery, and separates into layers; also, off-flavours develop during freezer storage. Slow freezing is especially harmful, as large ice crystals leave large holes, and the water frozen out is not reabsorbed during thawing, leading to a high drip loss.

Starch. Retrogradation of amylose results in a separation of liquid upon thawing, a phenomenon which influences the stability of frozen products, such as sauces, desserts, and cooked foods thickened with starchy agents; the use of amylopectin (waxy flour) or modified starches as thickening agents result in products much more stable to frozen storage.

Pectin. The enzyme pectinesterase may cause 'cloud loss' after reconstitution of frozen concentrated citrus juice.

Other constituents. Delay in the processing of peas and some other vegetables results in changes in flavour and content of vitamin C, and also increases toughness. The chemistry of the texture deterioration is unknown.

Appearance

Any change in appearance from the natural (fresh) character is regarded as a quality defect, and in many green vegetables, the change of chlorophyll to pheophytin is a well-known process due to simple hydrolysis. The bright green colour becomes more dull and yellow, especially at temperatures warmer than $-18°C$; proper blanching (time and temperature) and proper storage temperatures minimise this quality defect.

Loss of natural pigments in fruits during frozen storage is not a serious problem, although anthocyanase or phenolase may cause degradation of anthocyanin pigments during thawing of red fruits. For frozen fruits packed in sugar syrup, the red colour may diffuse into the syrup. This happens very slowly at $-18°C$ or colder, but can be very fast at warmer conditions.

The colour of frozen beef often deteriorates during frozen storage. In fresh beef, the myoglobin is found in the oxygenated form, oxymyoglobin, which is bright red. In vacuum-packaged beef, oxymyoglobin is reduced to purple myoglobin. During frozen storage, myoglobin (or oxymyoglobin) is oxidised to the brown metmyoglobin, presumably a photo-oxidation (MacDougall, 1982). It has been shown (Taylor, 1979) that the bright red colour in beef is stable for months at $-20°C$, depending on the permeability of the packaging, but after a few days in an open display cabinet with 'normal' illumination, the attractive colour has faded and discoloration (metmyoglobin) has appeared.

Formation of off-colour. Browning of frozen vegetable products, e.g. peaches, apples, mushrooms and potatoes, is due mainly to the interaction of an enzyme, polyphenol oxidase (or phenolase) with oxygen and colourless phenolic substances, resulting in the formation of brown polymers. The reaction can be prevented by adding reducing agents, such as ascorbic acid (vitamin C) or sulphur dioxide, if blanching is not appropriate. A packaging method which excludes oxygen also prevents this reaction. Frozen strawberries may change colour from bright red to reddish grey at temperatures warmer than $-18°C$. Polyphenoloxidase is absent in strawberries, but the reaction is still presumably an oxidation. Some vegetables develop off-colours, especially when stored at temperatures warmer than $-18°C$. In some cases, e.g. Brussels sprouts, a severe blanching could prevent this, but this would cause excessive loss of chlorophyll. Similarly, cauliflower may go dark, and in both cases the reaction is unknown.

Flavour

Flavour and odour changes occur during the frozen storage of most foodstuffs, especially in foods with a high lipid content.

Rancidity is the result of oxidation (autoxidation) of polyunsaturated fatty acids, and, through a group of reactions, a range of organic compounds are produced, including saturated and unsaturated aldehydes, ketones and alcohols. Aldehydes are the main cause of off-flavour and odour, and they may also react with amino acids (Maillard reactions) to form coloured compounds. This is seen in frozen fatty fish, where rancidity is accompanied by yellow to brown discoloration of the exposed surfaces of the fish. Rancidity is accelerated by freezer burn, as the fat is no longer protected by a barrier of ice, and also by pro-oxidants, such as trace metal ions (copper, iron), haemoglobin, and salt (NaCl). Rancidity is retarded or prevented by adding antioxidants, or by proper packaging, e.g. vacuum packaging in a tight-fitting material with low oxygen permeability. Low temperature storage retards rancidity; however, for cured meat, unpacked or packed in plastic materials with a high oxygen permeability, colder storage temperatures may increase rancidity, probably due to the cryo-concentration effect.

Off-flavour and odour in foods with a low content of lipids is caused by relatively unknown reactions; as vacuum-packaging retards the formation of off-flavour, oxidation is undoubtedly involved. In unblanched vegetables, off-flavours develop rather quickly. The cause may be rancidity, due either to oxidation of unsaturated fatty acids (the enzyme lipoxygenase) or hydrolysis of lipids (lipase). These enzymes are inactivated by proper blanching, and, in general, the chemistry of formation of off-flavours in frozen fruits and vegetables is not well understood.

Nutritional Status of Frozen Foods

As mentioned above, freezing preservation may often be the best way of preserving the quality, including the nutritive value, of foods which have to be stored for some time. Freezing and thawing hardly changes any nutrient, and although protein denaturation may occur, measurements of protein value have not resulted in any measurable reduction in the nutritional value of the protein. Similarly, lipid oxidation is often found in frozen foods, especially in animal products. However, foods with severe rancidity are not likely to be consumed, and the possible slight rancidity and oxidation, which may take place in some frozen foods, would have no measurable effect on the nutritive value of the lipids. Drip loss results in some losses of water-soluble vitamins and minerals, but the recorded losses are normally small.

It seems that the most important nutritional change is the loss of vitamins, especially the more sensitive vitamins such as B_1, B_2, and C. However, it must be emphasised that, when comparing the nutritive value of fresh and frozen foods, one has to measure the nutritive value

TABLE VI
The loss of vitamin C. After Boegh-Soerensen *et al.*
(1981)

	Loss of vitamin C (%)	
	Fresh peas	*Frozen peas*
Blanching	—	25
Freezing	—	0
Thawing	—	4
Home cooking	56	32
Total	56	61

of the food as it is eaten, and the data should also indicate the percentage of nutrient retained from the original product; otherwise, variations in cooking yields may lead to misleading results. Following this method of calculation, and where losses in home cooking are taken into consideration, results could be as shown in Table VI. From this Table, it can be seen that frozen peas, when purchased, have a lower vitamin C content than unfrozen peas, and that this is due mainly to losses during the blanching process which precedes freezing. This reduction is offset by a smaller loss during home cooking, making the difference between the frozen and the unfrozen product very small. The situation is often more favourable for frozen vegetables as some—more or less effectively chilled—fresh vegetables may quickly lose a significant percentage of vitamin C.

On the other hand, the blanching of other vegetables may result in vitamin C losses which exceed those in reasonable home cooking; much depends on the methods used in blanching and in the final cooking. Changes in the content of vitamin C in fruits and vegetables are small if the storage takes place at or below −18°C. At −30°C, there are no losses during 12 months, but at temperatures warmer than −18°C losses will occur, for example, for strawberries, a loss of about 50% after 8 months at −12°C.

In meat and meat products, the losses of vitamins are dependent on the type of product. However, the results from different authors on losses of B-vitamins during freezer storage are very conflicting. Mikkelsen *et al.* (1984) found very little difference between storage at −12°C and at −24°C. These authors concluded that storage for one year resulted in losses of thiamin of around 20 per cent (from 10 to 50 per cent); after one year of frozen storage, losses of riboflavin in meat varied from 44 to −10 per cent (i.e. a 10 per cent gain) and for pyridoxins from 52 to −33 per cent (i.e. a 33 per cent gain), and in both cases, broadly independent of storage temperatures between −12°C and −24°C. In general, the losses of riboflavin and pyridoxin are a little smaller than the losses of thiamin. The remaining B-vitamins are more stable than thiamin. For poultry and fish, the losses of vitamins are about the same as in meat.

Meat and Poultry

Large quantities of meat and poultry are frozen. Retail-packed frozen meat has not gained a substantial market share, whereas most of the poultry sold, especially chicken, is frozen.

Processes in meat

At slaughter, glycogen is converted to lactic acid causing a fall in muscle pH from 7 in the live animal to an ultimate pH of 5·5–6 in meat, and 5·9–6·4 in poultry. At the same time, the concentration of ATP in the muscles decreases, leading to rigor mortis. Rigor mortis is developed in about 10–30 h for beef, 8–16 h for lamb, 4–8 h for pigs, and 2–4 h for chicken.

DFD. If an animal is exhausted at the time of slaughter, the glycogen reserves are small, and the ultimate pH will be higher than normal. An ultimate pH above 6.4 may result in so-called DFD meat, i.e. meat which is dark, firm and dry. DFD meat has a high water-binding capacity, and although the storage life of chilled (vacuum-packed) DFD meat is reduced, there seem to be no special problems in the freezing preservation of DFD meat.

PSE. The problem of PSE meat (pale, soft, exudative) is confined almost entirely to pigs. There seems to be no special problem in the freezing of PSE meat, although the drip loss is, presumably, higher than normal.

Cold-shortening. If the temperature of a hot carcass is lowered too rapidly, a phenomenon known as 'cold-shortening' occurs, especially for beef, veal and lamb. This irreversible process may cause considerable toughness in the meat after cooking, and it will occur if the meat temperature is reduced below 10°C before the pH has fallen below about 6.2. In practical terms, the temperature in any part of beef, veal or lamb should not be permitted to fall below 10°C within 10 h of slaughter. Cold-shortening is not normally encountered in broilers, nor in pigs. The procedure of controlled chilling to avoid cold-shortening is called conditioning, and because the freezing of beef, veal and lamb should not be commenced too early after slaughter, direct freezing (one-stage freezing) is seldom used. Cold-shortening may be avoided by electrical stimulation, and this is used commercially for beef and lamb in some countries.

Thaw rigor. A phenomenon related to cold-shortening is thaw rigor. If meat is frozen quickly before rigor, and then is rapidly thawed, it contracts violently, exudes excessive drip, and becomes tough. Thaw rigor can be prevented by ensuring that the meat is completely in rigor before freezing, i.e. the meat should be chilled before freezing, or electrical stimulation should be applied. It seems that thaw rigor becomes less serious after some months in freezer storage, and that a slow thawing could reduce the deleterious effects.

Ageing (also called ripening or maturing) is used to increase the tenderness of meat, and the ageing time increases with decreasing temperatures. At 4°C, ageing may require 2–3 weeks for beef, one week for veal, 4 days for lamb, 2–3 days for pigs, and a few hours for chicken. For cuts of beef, veal and lamb, an ageing period is often used before freezing to ensure tender meat. However, ageing may reduce the shelf-life of the frozen products.

Meat

Meat may be frozen as carcasses, as packaged primal or consumer cuts, or after comminution and processing. The influence of freezing and thawing is mainly an increase in drip loss. The influence of freezing rate and/or thawing rate has long been a source of conflict, as summarised by James *et al.* (1984). These authors found no clear relation between freezing and percentage drip loss, but found a tendency for very fast thawing (microwave thawing) and very slow thawing (30–40 h) to produce greater than average losses. They concluded that there are considerable differences between muscles in the same carcass, and between areas within the same muscle, because of differences in collagen and fat content, and differences in temperature/pH history (during chilling).

Changes during freezer storage. So-called protein denaturation is probably the reason for meat becoming more dry during freezer storage. Processes leading to changes in taste and aroma are presumably of an enzymatic nature, but, apart from rancidity, very little is known about these processes. Desiccation (freezer burn) is a problem, as carcasses (sides, quarters) are normally unwrapped; it is difficult to wrap large carcasses, and their shape makes tight-fitting packaging difficult. Nevertheless, as shown in Table VII (p. 77), the practical storage life (PSL) of meat is comparatively long. The PSL for pork is somewhat shorter than that for beef and veal because there are more rancidity problems with pork as it contains more unsaturated fatty acids.

Poultry

Domestic poultry, e.g. chickens, turkeys, ducks and geese, is very frequently preserved by freezing, and is frozen as carcasses, as cuts (parts), or after preparation. The drip loss from frozen chicken is lower than for red meat, although the final result depends on the water chilling normally used in production. It is often discussed whether an ageing period prior to freezing would improve product taste and texture, and in some countries, turkeys are aged 12–24 h at about 0°C; chickens are seldom aged. If turkeys are not aged before freezing, an ageing period after (included in) the thawing process often improves the texture.

As mentioned earlier, a light (whitish) appearance of frozen poultry is often desired. Thus, packaged turkeys are sometimes crust frozen by passage through a glycol or brine spray for about 15–35 min. Apart from such initial freezing, poultry is practically always air-blast frozen, normally in cartons in which the plastic packaged birds are placed. After freezing, the cartons are lidded and strapped or glued. Freezing rate has little influence on the eating quality of poultry.

Poultry, and especially broilers where rancidity problems are negligible, have a long shelf-life, even at comparatively warm freezer storage temperatures, i.e. $-10°C$ to $-12°C$ (see Table VII). The processes leading to quality deterioration are of the same nature as described for meat.

Fish

Freezing preservation is very important because fish deteriorate more rapidly than most other foodstuffs, and they are often caught during a short season, frequently far away from their place of consumption. The body temperature of fish is close to the temperature of the fishing water, meaning that chilling (especially regarding the temperate regions) is not as effective as for meat. Freezing is used on the fishing vessel, or ashore, for subsequent thawing and further processing, or for the production of finished products.

Processes in fish

Fish muscle is similar to that of meat animals, although the muscle fibres are organised somewhat differently giving fish a more 'flaky' structure. Post-mortem glycolysis follows much the same path as in meat, but one difference between fish and meat is the generally lower glycogen content in fish, because the methods used for catching fish generally lead to exhaustion before death. Therefore, the fall in pH is smaller in fish, and the ultimate pH is seldom lower than 6·4–6·8. Fish flesh with a low ultimate pH will have more of a tendency to gape, and to develop toughness during frozen storage. The time taken for fish to enter rigor mortis can be anything from a few minutes to several hours; it varies with species, with degree of exhaustion at death, with nutritional status, and with temperature.

Ageing is not used for fish as the quality of fish—except for a few species—will not improve, and cold-shortening, except perhaps in tropical species, is not a problem. However, similar reactions may lead to contraction of fish muscle at high temperatures, especially when the fish temperature is higher than about 17°C. This may lead to gaping, and hence fish should be chilled as quickly as possible. Thaw rigor is seldom a problem with fish.

Changes during freezer storage

Fish lipids contain a much higher proportion of polyunsaturated fatty acids than meat lipids, and frozen fish are, therefore, very prone to rancidity. This is particularly noticeable in fatty fish, although lipid oxidation can also occur in lean (and flat) fish. Lipid content and composition varies within species with time of year, catching ground, and other factors, and the lipids markedly influence the storage life of the frozen product; fish with a low lipid content generally have a longer PSL than have fatty fish. Fish with a high oil content are the most susceptible to oxidative flavour and colour changes, but the type of oil is also important, as well as the presence in fish of natural antioxidants (e.g. tocopherols) and pro-oxidants (e.g. haem compounds and salt). White fish, i.e. fish with a low fat content, develop flavours and odours at rates which depend on storage time and temperature, and this phenomenon is sometimes described as 'cold storage flavour'. One of the compounds involved has been identified as hept-*cis*-4-enal (McGill *et al.*, 1974).

The proteins of fish also differ from those of meat, and frozen storage results in increased drip loss, and increased toughness and dryness on cooking. This change is due to denaturation and cross-linking reactions of the myofibrillar proteins, and can be detected in the laboratory by a progressive reduction in protein extractability in salt solutions (about 5% salt). These protein changes seem to be connected with the accumulation of free fatty acids—the result of lipases and phospholipases—or formaldehyde—the result of the cleavage of trimethylamine-oxide (TMAO) to dimethylamine (DMA) and formaldehyde.

All frozen products benefit from colder storage temperatures, and a temperature of −28°C to −30°C is generally recommended for long-term storage of whole fish, fish blocks, and fish products. Glazing, i.e. giving a protective coating of ice by dipping the frozen product in water, has been used for several years to protect fish against dehydration and oxidation. Now, glazing is often combined with, or replaced by, proper packaging.

Fruits and Vegetables

Products of vegetable origin behave differently from animal products during frozen handling in many ways, and of particular note is the fact that the cell walls are rich in cellulose, and are more rigid than the cell walls in animal tissues. The cell membranes, however, are very fragile, and, during industrial freezing and thawing, they become very permeable and allow the free movement of water, salts, carbohydrates and pigments; this leads to a drip loss, the quantity of which depends on several factors. The loss of turgor—internal pressure—on thawing, attributed to the collapse of the individual cells, is a special problem with fruits, e.g. strawberries and raspberries, but also with a number of vegetables, such as tomatoes, lettuce, cucumber and radish, i.e. products that are normally eaten in the raw state. The loss of turgor makes it impossible to freeze satisfactorily products that are to be consumed raw, but for most vegetables (peas, spinach, beans, Brussels sprouts, cauliflower, etc.), the frozen product exhibits a very good quality after thawing and cooking. The variety selected plays an important role as some varieties are much better suited to industrial preparation and freezing than others. Considerable care is, therefore, necessary to select the right variety, and having done so, to ensure that it is harvested at or near optimum maturity, which may be a matter of a few hours, e.g. for peas. Another important point is that vegetables may rapidly lose sensory and nutritive value after harvesting. At about 20°C, spinach could lose 75% of its vitamin C content in 24 h, and for a number of vegetables, flavour, colour and texture suffer if more than 2–3 h elapse before freezing (Neumann *et al.*, 1967). For fruits, it may be important to harvest at the right degree of ripeness, and then to quickly cool or freeze, especially delicate products with a high respiration rate.

Washing of vegetables is indispensable, and operations like sorting, grading, peeling and cutting should be carried out under hygienic conditions. These operations are normally performed mechanically, as are the procedures for peeling, stoning, coring, and slicing fruits; for peaches and apricots, these processes may be postponed until after thawing.

In vegetables, there is a rather high level of enzyme activity, which is delayed, but not stopped at freezer storage temperatures, and unless these enzymes are inactivated, the majority of vegetables will become unmarketable rather quickly, even if stored at temperatures colder than −18°C. It is for this reason that the blanching of vegetables has become general practice. Fruits are not normally blanched, but additives like ascorbic acid are used to inhibit an enzymatic browning which might otherwise occur.

Blanching is carried out by heating the vegetables in water, normally at 90–98°C, or in steam at about 100°C, for a relatively short time; usually 1–10 min depending on several factors, such as the dimensions of the vegetable, its nature and its structure, size and state of maturity. Grading is important to ensure uniform blanching, especially for Brussels sprouts, cauliflower and carrots.

Blanching inactivates enzymes, but it also has other advantages, in that it kills some of the micro-organisms on the plant surface, reduces the undesirable flavour of some vegetables, and helps to expel air from the tissues. Measuring the residual activity of certain enzymes is often used as an indicator of the adequacy of the blanching process. Tests for peroxidase and catalase are most often used, but tests for polyphenol oxidase or lipoxygenase are used in some cases. However, there are no official methods for measuring residual enzymatic activity, and little knowledge about the optimum enzymatic activity that should be present after blanching (and cooling). Thus, peroxidase is rather heat stable, and normally the blanching process should not completely inactivate this enzyme, as this would indicate too severe a heat treatment. For some vegetables, either the enzymatic activity is not high enough to necessitate blanching, or the organoleptic quality suffers so much on blanching that the process has to be omitted, the latter is especially true for herbs and aromatic spices (Philippon, 1981). In some cases, vacuum packaging might ensure an adequate storage life, but in others, storage at very cold temperatures is advisable. Regeneration of enzymatic activity, especially peroxidase activity, is sometimes observed during frozen storage of blanched vegetables, but the degree of regeneration is so low that it seems to have no effect on the quality of the food.

Cooling should be carried out immediately after blanching, and must lower the temperature so quickly that any detrimental effect of blanching, and the growth of micro-organisms, is limited. The most common cooling method is immersion cooling in cold water, but spraying with cold water and air-blast cooling is also used. The cooling process is a rather critical stage, as vitamins, minerals, and other soluble substances may be removed from the product, so creating product loss, and also pollution of the effluent.

After drainage, the foodstuffs are frozen immediately, often in a fluidised-bed freezer, and bulk packaged until retail packaging is carried out. In other cases, the product is packed into retail packs before freezing. The advantages of fluidised-bed freezing are that it is a rapid freezing method, and that, being an in-line freezer, its use ensures a very short delay between blanching and freezing. However, although the quality of most vegetables will not improve with ultra-rapid freezing, the quality of most fruits is enhanced in terms of decreased drip loss, and reduced loss of turgidity; even the most rapid commercial freezing will still result in the thawed fruit being much softer than the unfrozen. There is no need, of course, for rapid freezing of berries intended for further processing. On the contrary, the freezing of berries in sugar is sometimes carried out slowly to permit the sugar to penetrate into the berries.

Products frozen on a fluidised bed are IQF (individual quick frozen) products, which permits the consumer to pour from a retail pack the desired amount. If temperature fluctuations in the freezer chain occur, and especially if temperatures are warmer than −18°C, the IQF products may 'melt together' because of inside frost formation. This agglomeration tends to reduce their appeal to the consumer, as do chemical reactions which may lead to changes in colour and flavour. The practical storage lives (PSL) for various fruits and vegetables are given in Table VII.

Thawed Foods and Refreezing

In a number of countries, thawed foods (meats) sold as chilled foods shall be labelled, for instance, 'Previously frozen—do not refreeze' or 'Thawed—use immediately'. This indicates that thawed products are looked upon with considerable suspicion, especially from the legislative point of view. However, actual practice, and a number of experiments, have shown that, given correct thawing, the keeping quality is the same for the thawed product as for the unfrozen product. For thawed fruits and vegetables, the keeping quality may be reduced because of the collapse of cells, which may favour enzymes and micro-organisms.

Refreezing is looked upon with even more suspicion. In real life, fish is very often frozen at sea, for instance in blocks, then tempered or thawed after freezer storage, cut up into fillets or used for frozen fish sticks or other frozen fish products. This has been done for several years, but refreezing of fish can reduce the quality and/or the storage life (MacCallum, quoted by Andersen *et al.*, 1965). In the meat industry, too, refreezing is often used; in preparing frozen meat products (e.g. hamburgers), a certain proportion of the meat is usually frozen. Simonsen (1974) froze, thawed and refroze broilers up to four times without obvious effect, and the taste panel even indicated a slight texture improvement. The conclusion is that refreezing is an absolutely safe procedure, provided that the thawing process is carried out correctly, the time from thawing to refreezing is minimised, the freezing process is not too slow, and the temperature of the thawed food is such that microbial growth is suppressed.

STORAGE LIFE OF FROZEN FOODS

Most frozen foods are stored for some time, and, during this period, gradual and irreversible changes will occur leading to a deterioration in quality. In some frozen foods, this quality degradation proceeds very slowly, so such products have a long shelf-life, but in this context, the quality of a foodstuff is difficult to define. The best definition may be that 'quality is what the consumer expects of the product'. These expectations depend on a number of factors including price, appearance, taste, texture, and overall impression. For the consumer, it is often important that it is easy to prepare the frozen product. Small-sized products (meatballs, peas, shrimps, etc.) are often IQF products, thus making it easy for the consumer to pour out the necessary quantity. However, if the product has been exposed to rather warm temperatures, and especially to fluctuating temperatures warmer than −18°C, the product will freeze together into a solid block and a certain amount of inside frost will be formed, so that, if the product is packed in a transparent plastic pouch, the consumer will hesitate to buy. If packed in a paperboard carton or non-transparent plastic pouch, the consumer, after opening the package at home, will probably not purchase that brand again. Desiccation, which could mean white spots or freezer burn, will also have a negative influence on the consumer, and this is often found when the package is torn or holes have been formed. Warm and fluctuating storage temperatures may also result in colour changes, e.g. in green vegetables. Taste, texture, and overall impression are evaluated when the product is eaten in the home, and, as discussed earlier, freezer storage will result in changes in taste, texture, and overall impression (which includes other factors, such as colour/appearance of the prepared product).

Quality Evaluation

The aim of the (frozen) food industry is to manufacture and distribute foods which the consumer will buy. The final quality evaluation takes place in the home, but the industry needs to carry out some quality evaluation, partly as a 'normal' facet of quality control, and partly to be able to determine the shelf-life of the product. The latter is rather important in relation to date marking—in some countries, 'open' date marking is used, but in others, a code (e.g. a 'star' system) is available—but in either case, satisfactory methods of testing are essential.

In industry and in research/regulatory institutes, quality evaluation is carried out by subjective (sensorial, organoleptic) methods or by objective methods. Subjective tests are carried out by means of a number of tasters in a panel, while objective tests use chemical or physical methods, e.g. chemical analysis for chlorophyll, vitamin C or peroxides, and physical measurement of texture (the tenderometer for peas). The correlation between subjective and objective methods is often rather bad, and as subjective tests are closest to the consumer's evaluation of the product, the subjective tests are indispensable.

TTT–PPP

TTT (time–temperature tolerance)

TTT is a frozen product's storage life at different storage temperatures. The first real TTT experiments were carried out in Albany, California, from 1948 to about 1960, and, apart from research in irradiation, this is by far the most comprehensive research in the field of food science and technology. In these experiments, the storage life at different storage temperatures was determined for a number of frozen foods, especially fruits and vegetables. Two types of storage life were defined: high quality life (HQL) and practical storage life (PSL).

HQL, originally denominated JND (just noticeable difference), is the storage time, at a given temperature, until a trained taste panel can detect a difference in comparison with a control sample kept at a very cold temperature. HQL is of interest in research, but not often in practice.

PSL is the time the frozen product can be kept at a given temperature, and still being fully acceptable to the consumer. PSL will often be about twice HQL, perhaps more.

The relationship between storage temperatures and storage life is often shown in a TTT-curve; Fig. 9 depicts the PSL for poultry parts. The influence of storage temperature on storage life is normally that colder temperatures increase storage life.

PPP (product–processing–packaging)

The so-called PPP-factors (product–processing–packaging) also have a very great influence on storage life.

Product, i.e. the nature and quality of the raw material, is very important. It is well-known that frozen beef keeps longer than frozen pork, and also that lean and flat fish keep longer than fatty fish, due to

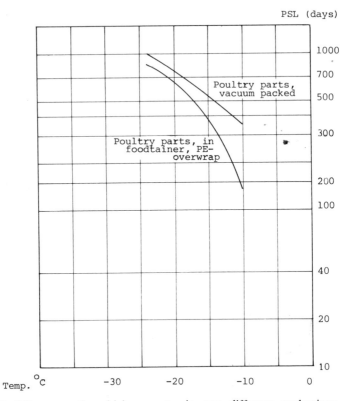

Fig. 9. PSL-curves for chicken parts, in two different packagings. After Boegh-Soerensen (1975).

their different lipid composition or a lower amount of lipids. Calf liver of good bacteriological quality has a HQL four times that of a similar liver with a high bacterial count. Similarly, aged beef will have a shorter shelf-life than unaged beef, presumably due to the action of bacterial enzymes.

Processing, i.e. the pre-treatment prior to freezer storage, often influences storage life. Heat processing, e.g. blanching of vegetables, results in a longer PSL for most products. Breading prolongs PSL, and the same is often the case with smoking. Hanson *et al.* (1959), however, found that frying chicken resulted in a very short PSL.

Packaging. It is now normal to use comparatively good packaging for frozen foods. A suitable package has a low permeability to water vapour and oxygen, and it fits the product tightly. In addition, the packaging material should be able to withstand temperatures down to about $-30°C$, and even $-40°C$ in some cases. Vacuum-packaging, using a suitable packaging material, is often the best package, but it is more expensive and, for some products, results in only small improvements in PSL.

Storage Life Tables

As the PPP-factors have an enormous influence on storage life, it is very difficult to work out precise tables for practical storage life (PSL). However, the PSLs for different frozen products are given at three temperatures ($-12°C$, $-18°C$, and $-24°C$) in Table VII, mainly to give an indication of the PSLs for different frozen products which have been processed and packaged according to good manufacturing practice. It must be re-emphasised that differences in the PPP-factors could affect the times indicated greatly.

TABLE VII
Practical storage life (PSL) in months for different frozen foods

Product	Storage temperature		
	−12°C	*−18°C*	*−24°C*
Beef cuts	8	18	24
Minced beef	6	10	15
Pork cuts	6	10	15
Chickens	9	18	30
Ducks, geese	6	12	15
Fatty fish	3	5	9
Lean fish	4	9	18
Carrots	10	18	28
Peas	6	24	30
Strawberries	6	15	24
Fruit concentrates	12	24	30

Loss of Storage Life in the Freezer Chain

Frozen foods spend the time from freezing until cooking in the freezer chain, i.e. producers' and wholesalers' freezer storage rooms, transport and distribution, retail shops (back-up storage rooms and frozen food

TABLE VIII
Loss of storage life for pork chops in a supposed freezer chain

Period of storage	Storage temperature (°C)	PSL (days)	Storage time (days)	Loss in PSL (%)
Producer	−25	500	120	24·0
Transport	−18	320	1·5	0·5
Wholesaler	−20	360	30	8·3
Retailer	−15	270	20	7·4
Transport	−10	120	0·1	0·1
Home freezer	−18	320	50	15·6
Total loss of storage life:				55·9

display cabinets) and home freezers. During this period, the quality of the frozen product will decline, with the rate of quality degradation depending on the storage times and the temperatures to which the product is exposed. If the time–temperature history is known or can be estimated, the quality loss—or more correctly the loss of storage life—that the product has experienced can be calculated. Table VIII shows such a calculation. Frozen pork chops in good packaging, with PSLs as indicated at the different storage temperatures, pass through an estimated freezer chain. The calculations are then very simple, as it is supposed that the daily loss in PSL is the same regardless of sequence (the rule of additivity); this is one of the conclusions of the above-mentioned Albany TTT experiments. For example, in the producer's freezer storage room, the mean temperature is −25°C; at −25°C, the PSL of pork chops is 500 days, i.e. a daily loss of 100/500 = 0.2 per cent; 120 days at −25°C means a 24 per cent loss of PSL. The total loss in PSL throughout the chain is 56 per cent; but as 44 per cent remains, the product could be stored at −18°C for 320×44/100 = 141 more days, and still be fully acceptable to the consumer. Such calculations have been shown (Bech-Jacobsen and Boegh-Soerensen, 1984) to lead to results which can be confirmed by taste evaluation; thus, the method is applicable in practice.

Table VIII shows that the effect of transport on quality loss is normally negligible. The weakest link in the freezer chain is the display cabinets. Open-top cabinets are most widely used, and, in such cabinets, product temperatures in the upper layer are very often warmer than −15°C, although the circulating cold air just above the

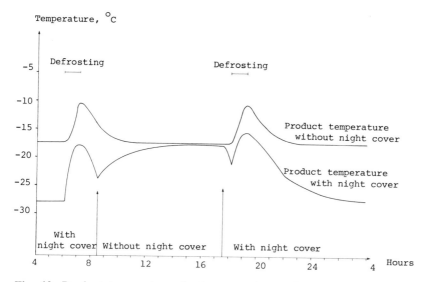

Fig. 10. Product temperatures in the upper layer of a frozen food display cabinet during 24 h, with and without night cover. (Boegh-Soerensen *et al.*, 1978).

upper layer is −25°C or colder (Boegh-Soerensen and Bramsnaes, 1977); product temperatures as warm as −3°C have been recorded. During defrosting (often two times a day), product temperatures in the upper and outer layers increase by 4–8°C, and this elevation may last 2–3 h before the temperature is reduced again to the level before defrosting. The use of night-covers reduces the average temperature, and although it results in greater temperature fluctuations, (see Fig. 10), it seems that the reduction in average temperature is decisive, i.e. the use of night-covers reduces the quality loss during residence time in open top display cabinets (Brinch *et al.*, 1983). At the warm and fluctuating temperatures in the upper layer, the 'normal' temperature may be −12°C, but with two defrosting periods per day, the effective mean temperature is about −8°C. At −8°C, the PSL of the pork chops in Table VIII is about 70 days, the quality loss about 28·6% in 20 days, and the total loss 77·1% instead of 55·9%. As some products spend considerably longer in display cabinets—although not all the time in the upper layers—the quality loss may be much larger. In a Danish survey (Boegh-Soerensen and Bech-Jacobsen, 1984) it was found that 5% of all frozen products spend more than 10 weeks in a display cabinet, and 10 weeks at −8°C would mean that the limit for acceptability had been exceeded. Also, Table VII shows that some frozen foods have a considerably shorter shelf-life than the pork chops referred to in Table VIII.

REFERENCES

Andersen, E., Jul, M. and Riemann, H. (1965). *Industriel Levnedsmiddelkonservering* (Industrial food preservation), Teknisk Forlag, Copenhagen.
Åström, S. and Löndahl, G. (1969). *Kylteknisk Tidskrift*, **28** (6), 1–4.
Bailey, C. (1981). Industrial thawing methods. In: *Minutes of COST 91, Sub-Group 3, meeting in Gothenburg (April 1981)*. Danish Meat Products Laboratory, Copenhagen.
Bech-Jacobsen, K. and Boegh-Soerensen, L. (1984). TTT-studies of different retail packed pork products. In: *Thermal Processing and Quality of Foods* (Eds P. Zeuthen *et al.*) (Proceedings of COST 91 symposium in Athens), Applied Science Publishers, London.
Boegh-Soerensen, L. (1975). *Fleischwirtschaft*, **55** (11), 1587–92.
Boegh-Soerensen, L. and Bramsnaes, F. (1977). The effect of storage in retail cabinets on frozen foods. *Bull. Int. Inst. Refrig.* (Annexe 1977–1), 375–80.
Boegh-Soerensen, L. and Bech-Jacobsen, K. (1984). Total time for frozen foods in the freezer chain and in display cabinets. In: *Thermal Processing and Quality of Foods* (Eds P. Zeuthen *et al.*) (Proceedings of COST 91 symposium in Athens), Applied Science Publishers, London.
Boegh-Soerensen, L., Jul, M. and Hoejmark Jensen, J. (1978). *Konserveringsteknik* (Food preservation technology), Vol. 1, DSR, Copenhagen.
Boegh-Soerensen, L., Jul, M. and Hoejmark Jensen, J. (1981). *Konserveringsteknik* (Food preservation technology), Vol. 2, DSR, Copenhagen.
Brinch, J., Boegh-Soerensen, L. and Green, E. (1983). Frostdiske: produktkvalitet, energiforbrug (Retail freezer cabinets: product quality, energy consumption). *Scandinavian Refrigeration*, **3**, 113–15.
Clemmensen, J. and Zeuthen, P. (1974). Physical factors influencing the quality of unwrapped pork sides during freezing. In: *Proceedings from MRI symposium No. 3: Meat freezing—Why and How*, Meat Research Institute, Langford.
Fleming, A. K. (1974). The New Zealand approach to meat freezing. In: *Proceedings from MRI symposium No. 3: Meat freezing—Why and How*, Meat Research Institute, Langford.
Hanson, H. L., Fletcher, L. R. and Linewater, H. (1959). *Fd Technol.*, April, 221–4.
IIR (International Institute of Refrigeration) (1972). *Recommendations for the processing and handling of frozen foods*, IIR, Paris.
James, S. J., Nair, C. and Bailey, C. (1984). The effect of the rate of freezing and thawing on the drip loss from frozen beef. In: *Thermal Processing and*

Quality of Foods (Eds P. Zeuthen *et al.*) (Proceedings of COST 91 symposium in Athens), Applied Science Publishers, London.

Jason, A. C. (1974): Rapid thawing of foodstuffs. In: *Proceedings of IFST,* **7** (3), 146–57.

Kostaropoulos, A. E. (1979). *Zeitschrift für Lebensmittel-Technologie und Verfahrenstechnik,* **30** (4), 133–6.

Löndahl, G. and Nilsson, T. E. (1978). *Int. Journal of Refrigeration,* **1** (1), 53–6.

Lorentzen, G. (1964). *Bull. Int. Inst. Refrig.* (Annexe 1964–1), 39–46.

Love, R. M. (1968): Ice formation in frozen muscle. In: *Low Temperature Biology of Foodstuffs* (Eds J. Hawthorne and E. Rolfe), Pergamon Press, Oxford.

Luyet, B. (1968). Basic physical phenomena in the freezing and thawing of animal and plant tissues. In: *The Freezing Preservation of Foods* (Eds D. K. Tressler, W. B. van Arsdel and M. J. Copley), AVI Publishing Company, Westport, Connecticut.

MacDougall, D. B. (1982). Changes in the colour and opacity of meat. *Food Chemistry,* **9,** 75–88.

McGill, A. S., Hardy, R. and Burt, J. R. (1974). *J. Sci. Fd Agric.,* **25,** 1477–89.

Mellor, J. D. (1978). *Bull. Int. Inst. Refrig.,* **3,** 569–84.

Mellor, J. D. and Seppings, A. H. (1976). Thermophysical data for designing a refrigerated food chain. *Bull. Int. Inst. Refrig.* (Annexe 1976-1), 349–59.

Mikkelsen, K., Rasmussen, E. L. and Zinck, O. (1984). Retention of Vitamin B_1, B_2 and B_6 in frozen meats. In: *Thermal Processing and Quality of Foods* (Eds P. Zeuthen *et al.*) (Proceedings of COST 91 symposium in Athens), Applied Science Publishers, London.

Miles, C. A. and Morley, M. J. (1977). *Journal of Food Technology,* **12,** 387–402.

Morley, M. J. (1972). *Thermal properties of meat: tabulated data,* Meat Research Institute, Langford.

Neumann, H. J., Dietrich, W. C. and Guadagni, D. G. (1967). *Quick Frozen Foods Int.,* Dec. 1967, 101–3.

Philippon, J. (1981). In: *Minutes of COST 91, Sub-Group 3, meeting in Paris (November 1981),* Danish Meat Products Laboratory, Copenhagen.

Plank, R., Ehrenbaum, E. and Reuter, K. (1916). *Die Konservierung von Fischen durch das Gefrierverfahren* (Fish freezing), Zentral-Einkaufgesellschaft, Berlin.

Polley, S. L., Snyder, O. P. and Kotnour, P. (1980). *Fd Technol.,* Nov. 1980, 76–94.

Riedel, L. (1950), *Kältetechnik,* **2,** 195–202.

Riedel, L. (1959). *Kältetechnik,* **11,** 41–3.

Riedel, L. (1964). *Kältetechnik,* **16,** 363–66; **5,** 129–33.

Riedel, L. (1977). *Chem. Mikrobiol. Technol. Lebensm.,* **5,** 118–27.

Riedel, L. (1978). *Chem. Mikrobiol. Technol. Lebensm.,* **5,** 129–33.

Simonsen, B. (1974). The quality of thawed and refrozen broiler chickens, Danish Meat Products Laboratory, Copenhagen. (Mimeographed.)

Spiess, W. E. L. (1977). *Fleischwirtschaft*, **5**, 968–77.

Taylor, A. A. (1979). Skin packaging of frozen meat. *Food Manufacture*, **54**, 61–3.

van den Berg, L. (1968). Physiochemical changes in foods during freezing and subsequent storage. In: *Low Temperature Biology of Foodstuffs*, (Eds J. Hawthorne and E. Rolfe), Pergamon Press, Oxford.

Chapter 3

Response of Micro-organisms to Freeze–Thaw Stress

R. Davies

National College of Food Technology, University of Reading, UK

and

A. Obafemi

Department of Microbiology, University of Port Harcourt, Nigeria

INTRODUCTION

Freezing has been used as a means of preserving biological materials since relatively early times. The study of its effects on the survival of micro-organisms, however, probably dates back to the 18th century observation by Leeuwenhoek and Spallanzani of the 'resurrection of small animalcules from frozen pond water' (Calcott, 1978). Today, there are two major reasons for studying the effects of freezing on microbes: on the one hand, to minimise their survival and prevent their multiplication in perishable materials, such as foods, and, on the other hand, to preserve the microbial cells themselves as in the case of medically and industrially important cultures. Both approaches need a common understanding of the factors affecting cryosurvival, and of the mechanisms of freeze-inactivation of microbial cells.

As a system is progressively cooled and vegetative micro-organisms are held at temperatures below their minimum for growth, some loss of cell viability can be anticipated. This is exacerbated when exponential phase cells are transferred suddenly to significantly lower temperatures, particularly near 0°C, where they experience 'cold shock' (Ingram and Mackey, 1976). Further inactivation takes place during the actual freezing process, and yet further viability loss occurs during prolonged frozen storage. These effects can only be monitored in thawed cultures, and thus the nature both of the thawing conditions and of the detection methods must also be taken into consideration.

The factors affecting the survival of microbes are complex, interrelated and regrettably not always clearly specified or separated in the early literature. The subject, however, has been extensively reviewed (Mazur, 1966; Ray and Speck, 1973; Merryman, 1974; Ingram and Mackey, 1976; Macleod and Calcott, 1976; Calcott, 1978) and therefore only selected aspects will be considered here.

PHYSICAL CHANGES DURING FREEZING

As aqueous solutions are cooled from temperatures above 0°C, the solubility of the solutes may change slightly, but generally, with the exception of saturated solutions, the systems remain liquid until reaching their freezing point at some temperature below 0°C. For dilute, biological systems this may be at −1°C to −3°C depending on the nature and concentration of the solutes present (Ingram and Mackey, 1976). On holding at, or on further cooling below, this temperature either ice crystals begin to separate, or, in the absence of nucleation, the liquid becomes supercooled. Assuming that ice eventually begins to form, the dissolved solutes are concentrated in the remaining liquid. As the temperature is further reduced and more water is converted into ice, the solute concentration rises more and more in the unfrozen water portion with corresponding decreases in freezing point and water activity (a_w). The relationship between cooling temperature and water activity is represented in Fig. 1.

This concentration of solutes continues, as the temperature is lowered, until the eutectic point is reached, and the remaining solution then solidifies. The lowest temperature at which the solution remains liquid is referred to as the eutectic temperature. In the case of NaCl, for example, the eutectic temperature is −21·8°C (Merryman, 1966), and the concentration attained before solidification is approximately 5 M.

The size of the ice crystals formed during freezing is primarily dependent on freezing rate. Solutions frozen at ultra rapid rates (several thousand degrees centigrade per minute) contain ice crystals of sub-microscopic size which can only be demonstrated by allowing subsequent crystal growth and aggregation at higher temperatures of about −30°C. Slow freezing rates (typically 1°C min^{-1}) produce large crystals which can exceed the dimensions of microbial cells.

The concept of 'freezing rate' is difficult to define and measure. As

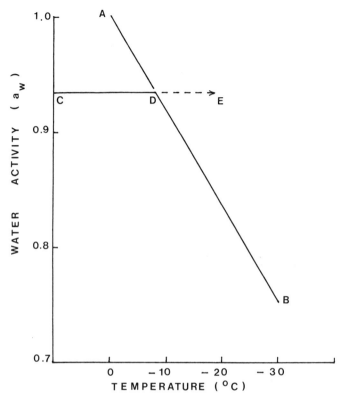

Fig. 1. The relationship of a_w to temperature for ice and water mixtures is represented by line AB. The effect of lowering temperature on the a_w of a high solute liquid is shown by line CDB where the freezing point is depressed to D, or alternatively if supercooling occurs the medium remains liquid to temperature E. (Similar diagram given by Ingram and Mackey, 1976.)

discussed by Merryman (1966), methods of quantifying freezing rates have tended to be inconsistent in the literature and have varied from specifying the length of time held in a freezing bath at fixed temperature, to attempting to force samples through their freezing points at specified rates. The difficulties are illustrated by the typical cooling profiles, obtained by the present authors, shown in Fig. 2. For each refrigerant freezing system, there are at least three recognisable stages in the temperature profile: first a rate of cooling to freezing point which is frequently non-linear; second a plateau at the melting point affected by the release of latent heat and passage of the freezing

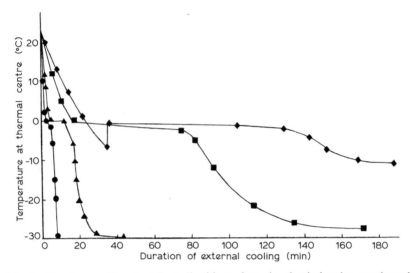

Fig. 2. Cooling profiles for 10 ml, liquid exudates in plastic bottles monitored by thermocouples. The freezing systems are: dry ice/acetone at $-78°C$ (●); liquid Freon at $-30°C$ (▲); chest freezer at $-23°C$ to $-34°C$ (■); freezer compartment of domestic refrigerator with 30-min cycles at $-9°C$ to $-19°C$ (◆). (Data of Obafemi, 1983.)

boundary and third, a rate of cooling after freezing which is largely dependent on thermal gradients and diffusivities. Merryman considers that the third stage correlates best with ultimate ice crystal size, and therefore he and the majority of subsequent workers (Calcott, 1978) specify freezing rates as 'the rate of temperature fall immediately following freezing measured over a temperature span where the temperatures fall is relatively linear'. In our own case we have adopted the approach of Takano et al. (1979), and measured the rate of cooling, after freezing, from $-5°C$ to $-10°C$. The freezing rates for dry ice/acetone, liquid Freon, chest freezer and domestic refrigerator–freezer systems are: 10, 2·5, 0·7 and 0·2°C min^{-1} respectively. Our value of 10°C min^{-1} for immersion in a dry ice/acetone bath, however, compares with a recently quoted value of 100°C min^{-1} (Ray, 1983) which demonstrates that inconsistencies still remain.

Microbial cells, suspended in aqueous solutions during freezing, behave like solute molecules and become partitioned and concentrated in the unfrozen portion of the solution as ice crystals form. They are thus exposed to the effects of concentrated solutes, and to the consequences of localised ice crystal growth (Calcott, 1978).

FACTORS AFFECTING MICROBIAL SURVIVAL

Nature of the Flora

Many psychrophilic and psychrotrophic micro-organisms are capable of growth at sub-zero temperatures (Larkin and Stokes, 1968). It appears that growth at temperatures down to $-5°C$ or $-7°C$ is fairly commonly observed, although only rarely at temperatures below $-10°C$. *Bacillus psychrophilus,* for example, grows at $-5°C$ to $-7°C$ with a generation time of 204 h (Larkin and Stokes, 1968). Generation times of this order, however, hardly provide opportunity for cell proliferation during freezing processes which, in most practical cases, are of relatively short duration. Differing sensitivity to 'cold shock', on the other hand, may be a more significant factor as it has long been considered that thermophiles and mesophiles were more susceptible to low temperature shock than psychrophiles (Jay, 1978). As pointed out by Ingram and Mackey (1976), however, cold shock effects depend more on the magnitude of any temperature drop rather than on the actual temperature at which the cells were grown; this has been demonstrated in psychrophiles (Farrell and Rose, 1968) as well as mesophiles, and in Gram-positive sporeformers and streptomycetes as well as Gram-negative bacteria. Nevertheless, since the mechanisms of cold shock damage seem potentially to resemble those of subsequent freezing injury (as discussed later), the interrelationships between the two stresses, for different types of organism, seem worthy of further investigation.

As freezing occurs and temperatures continue to drop, the associated progressive reduction of a_w has a selective effect on the microflora. Organisms which grow on frozen systems, provided that other environmental factors are supportive of growth, reflect a trend favouring dehydration resistant types such as yeasts and moulds rather than bacteria (Ingram and Mackey, 1976). Little regard seems to have been given to the relative freezing resistance of organisms which are tolerant of conditions of reduced a_w (e.g. osmophiles, halophiles, xerotolerant fungi), and again this is perhaps an area worthy of study. Generally, however, there is little evidence to indicate that an ability to grow in the conditions associated with freezing (i.e. in the sub-zero conditions of supercooling or in the reduced a_w conditions after ice formation) imparts any additional resistance to the physical stresses of freezing to either psychrotrophs (Elliott and Michener, 1965) or to yeasts and moulds (Ingram and Mackey, 1976).

Micro-organisms, therefore, can be grouped according to differences in their inherent responses to freeze–thaw stress by using the categories of Mazur (1966), that is organisms that:

  (a) survive all conditions of freezing and thawing;
  (b) resist the immediate effects of freezing but are sensitive to frozen storage;
  (c) are sensitive to both immediate and storage effects of freezing under some conditions;
  (d) are sensitive to freezing and frozen storage under all conditions.

The first category includes most spores. Bacterial endospores are extremely resistant to freezing and to storage at sub-zero temperatures, with survival levels exceeding 90%. This can be attributed to the relatively dehydrated state of the spore protoplast with much of its water bound in an unfreezable state within the expanded cortex (Gould and Dring, 1975). Fungal spores are also resistant, and Mazur has shown that air-dry conidia of *Aspergillus flavus*, cooled rapidly to −73°C and thawed rapidly, gave a survival level of 75%. This reduced to 3·2% for spores suspended in water before freezing, and to less than 1·0% for frozen spores thawed slowly. A few vegetative bacterial cells are also fairly insensitive to freezing. Some Gram-positive staphylococci, micrococci and streptococci are relatively resistant, with survival exceeding 50% (Mazur, 1966), and various authors have found enterococci to be resistant (Ingram and Mackey, 1976; Olson and Nottingham, 1980). Strain differences for lactic streptococci, however, have been attributed to differences in cellular contents of fatty acids and polyglucose compounds (Gilliland and Speck, 1974).

Organisms which are very sensitive to the effects of freezing, i.e. Mazur's category (d), include the free living amoebae, ciliated protozoa and nematodes. Storage at −10 to −20°C for a few days has been shown to be lethal to *Toxoplasma gondii*, *Entamoeba* and trypanosomes (Olson and Nottingham, 1980), though cooling rates are critical, and lethality seems to be associated with the formation of intracellular ice at rates of 3°C per minute and above (Calcott, 1978).

The majority of micro-organisms, however, are in categories (b) and (c). Generally, most Gram-positive organisms including *Bacillus, Clostridium, Corynebacterium, Lactobacillus, Microbacterium, Micro-coccus, Staphylococcus* and *Streptococcus,* together with some yeasts, are relatively resistant to freezing, though some, such as *Cl. perfringens,* are very sensitive to frozen storage (Trakulchang and Kraft, 1977; Olson and Nottingham, 1980). Gram-negative organisms, such as *Escherichia, Salmonella, Serratia, Pseudomonas Acinetobacter-Moraxella* and *Vibrio,* are considerably more sensitive to both freezing and frozen storage (Ingram and Mackey, 1976; Olson and Nottingham, 1980), and their survival is dependent on a number of variables including cooling velocity, temperature, menstruum, cell concentration, storage time and thawing conditions (Mazur, 1966). Typically, the microflora of raw meat, before freezing at $-30°C$, contained 15% Gram-positive bacteria and 85% Gram-negative. After freezing, when the total viable count fell from 385 000 to 77 000 g^{-1}, the proportions reversed to become 70% and 30% respectively (Partmann, 1975). Exponential phase cells of *Salmonella typhimurium,* frozen in distilled water at $-30°C$ by immersion in liquid Freon for 44 min, showed less than 0·1% survival after rapid thawing (Obafemi, 1983), thus illustrating the cryosensitivity of a significant, Gram-negative pathogen.

Nutritional Status

The influence of previous nutritional conditions of growth on the subsequent sensitivity of microbial cells to various stresses, including freezing, has been reviewed by Calcott (1977). Prompted by the speculations of Gilliland and Speck (1974) that it should be possible to manipulate the cryosurvival of streptococci by altering those growth conditions which promote the synthesis of protective polymers, Calcott and Macleod (1974*a,b*) have investigated this factor by growing cells of *Escherichia coli* under different carbon- or nitrogen-limitation conditions in continuous culture. They found that nitrogen-limited cells accumulated higher carbohydrate contents and were more resistant to freezing, and this observation led to the suggestion that polyglucose and glycogen-like reserve materials could be cryoprotective by strengthening the cell envelope or outer membrane. Obafemi (1983) found that exponential phase cells of *S. typhimurium* grown in tryptone soya broth were 100 times more sensitive to freezing at $-30°C$ than cells grown to exponential phase in a minimal medium. The reasons for this observation are not clear, but may be related to the nitrogen-limited status of the minimal salts–glucose medium which contains a total nitrogen content of approximately only 0·0003% w/v.

Age and Growth Rate

Exponential phase cells are much more sensitive to freeze–thaw stress than stationary phase cells, as illustrated for *S. typhimurium* in Fig. 3. This is a generally recognised phenomenon and has been widely reported (Toyokawa and Hillander, 1956; Johannsen, 1972; Ray and Speck, 1973; Calcott, 1978), yet many workers have used stationary phase cells for freeze–thaw studies (Ingram and Mackey, 1976), and

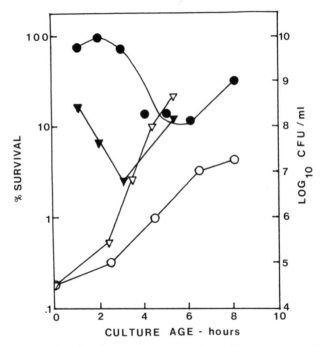

Fig. 3. The relative freezing resistance of *S. typhimurium* as a function of previous growth temperature and phase of growth. Cells were grown in chicken exudate (obtained by repeated freeze/thawing of comminuted chicken portions followed by membrane filtration) at 25°C (O) and 37°C (▽), harvested at different points on the growth curves and subjected to freeze–thaw stress (liquid Freon/40°C respectively). Survivors were enumerated on tryptone-soya-agar plates (closed symbols). (Data of Obafemi, 1983).

Packer *et al.* (1965) were not able to demonstrate a clear relationship, for *E. coli,* between freezing susceptibility and phase of growth. The mechanisms of growth-phase-related changes in cryosensitivity have not been subjected to definitive study, though it has been suggested, quite logically, that the differences merely reflect differences in physiological activity and hence vulnerability between the two states of growth (Ray and Speck, 1973). However, Davies (1970) found that the sensitivity of *Pseudomonas* to freezing, at different phases of growth, was dependent on freezing rate. At lower cooling rates of $10°C \, min^{-1}$ or below, exponential phase cells were more sensitive, but at higher rates of $100°C \, min^{-1}$ and beyond, stationary phase cells were more sensitive. This work was extended by Calcott and Macleod (1974*b*) using chemostat populations of *E. coli* grown at various rates. At low cooling rates, cells became more susceptible to freezing stress as their growth rates increased (and this was also true for cells frozen at ultra-rapid cooling rates), but at rates between $10°C \, min^{-1}$ and $100°C \, min^{-1}$, the relationships were sometimes reversed. Our own data (Fig. 3) for *S. typhimurium,* in addition to demonstrating growth-phase-dependent changes in freezing sensitivity, also show that cells grown at 25°C are more resistant to freezing stress than cells grown at 37°C. This contrast may involve changes in membrane properties as a function of growth temperature (Ellar, 1978), or reflect the different growth rates at mid-exponential phase at 37°C and 25°C which are $0.25 \, h^{-1}$ and $0.17 \, h^{-1}$ respectively.

Rate of Cooling

The effect of 'cooling rate' on the freeze-inactivation of microbial cells is fairly well documented (Mazur, 1966, 1970; Calcott and Macleod, 1974*a*; Macleod and Calcott, 1976; Calcott, 1978), and the response pattern is similar for many cell types. Figure 4 shows how the survival of *E. coli* during freezing to $-70°C$ depends on the cooling rate. The cells have an optimum for survival in the slow cooling rate range $(10°C \, min^{-1})$, but as the rate of cooling is increased or decreased, survival is reduced. At very high rates of cooling, viability increases to approach a maximum at ultra-rapid cooling rates $(10\,000°C \, min^{-1})$, though rates of this order would seldom be experienced in commercial practice (Ingram and Mackey, 1976). The more usual methods of freezing, ranging from domestic, two-star freezers to the use of liquid

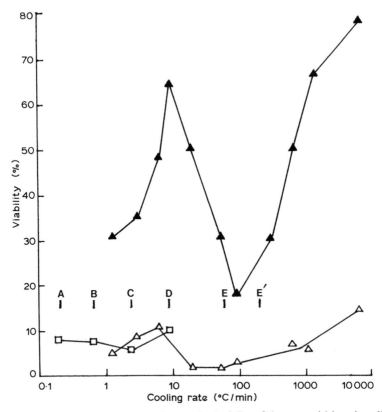

Fig. 4. Effect of cooling rate on the survival of *E. coli* in water (▲) or in saline
(△). (From the data of Calcott and Macleod, 1974*a,b*.) Effect on the survival
of *S. typhimurium* in chicken exudate (□) (Obafemi, 1983). A, B, C, and D
indicate the cooling rates for: domestic refrigerator–freezer, chest freezer, liquid
Freon immersion and dry ice/acetone immersion respectively (Obafemi, 1983).
E and E′ represent the range of cooling rates for liquid N_2 freezing (Souzu, 1980).

N_2, are associated with cooling rates from 0·2°C min^{-1} to 200°C min^{-1}
as identified in Fig. 4.

The existence of an optimum cooling rate, for survival in the low
cooling rate range, suggests that there are at least two mechanisms of
damage which are oppositely affected by rate of cooling. Mazur (1966)
suggested that at very slow cooling rates, below the optimum for
survival, extracellular freezing takes place and cell damage is mainly
caused by exposure to increasing concentrations of solutes. As cooling

rates increase, the exposure times to these solute stresses are reduced, with resultant increases in cell survival. At higher cooling rates ($>10°C\,min^{-1}$), however, intracellular ice begins to form and survival falls. The increase in survival at ultra-rapid cooling rates seems to be a function of the rapid warming rates used in the thawing procedures, and is markedly reduced when these are controlled to low values, e.g. $12°C\,min^{-1}$ (Macleod and Calcott, 1976).

Freezing in the presence of 0.85% NaCl causes a reduction in survival of *E. coli* (Fig. 4), but the response pattern still reflects a cooling-rate-dependent effect. Similarly the data for exponential phase cells of *S. typhimurium*, frozen in chicken exudate (obtained by repeated freeze/thawing of fresh chicken carcasses and containing 5% blood serum protein), also showed a slight rate-dependent effect.

Frozen Storage

Organisms lose viability when held at or just below freezing temperatures, but as the holding temperature is lowered, the death rate reduces, and below $-60°C$ the rate of death is usually very low or zero (Macleod and Calcott, 1976). After freezing, death is initially fairly rapid, particularly at $-2°C$ to $-5°C$, and then gradually slows on holding at a constant sub-zero temperature until eventually, in the late stages of storage, viable numbers remain almost constant. The decline in viability is probably because of continued exposure to concentrated solutes, and thus represents an extension of the immediate stresses associated with freezing at slow cooling rates (Ingram and Mackey, 1976). Storage death may also be affected by differential precipitation of solutes with time (van den Berg, 1968), by pH changes (Mazur, 1966) and by ice crystal growth due to recrystallisation (Davies, 1970).

Thawing

It is generally considered that the rate of thawing has very little effect on the survival of microbial cells which have been previously frozen at low cooling rates ($<100°C\,min^{-1}$) (Calcott, 1978). At rapid and ultra-rapid cooling rates, however, subsequent survival is considerably greater at rapid warming rates ($>500°C\,min^{-1}$) than after slow thawing ($<12°C\,min^{-1}$) (Mazur, 1966; Macleod and Calcott, 1976). This has been explained as a consequence of ice crystal growth during slow thawing which is prevented or minimised in conditions of rapid thaw (Ingram and Mackey, 1976).

In contrast, however, Obafemi (1983) has shown that the thawing of liquid-Freon-frozen, exponential phase cells of *S. typhimurium* at 4°C for 80 min is more lethal than thawing at 40°C for 13 min, with an attendant increase in the deoxycholate sensitive proportion of the survivors at 4°C. This effect was not observed for stationary phase cells where the degree of both lethal and reversible injury was minimal at both warming temperatures.

Enumeration of Survivors

Microbial viability after freeze–thaw stress is usually assessed by determining colony-forming ability by conventional plate or slide culture methods using non-selective media. As discussed by Calcott (1978), these methods are subjective, and the failure of organisms to produce colonies may result from the original freeze–thaw stress, or from further stresses or inadequacies associated with the enumeration media and conditions. In general, it is accepted that after freezing and thawing, more bacteria are detected in rich rather than poor media, and that energy-dependent repair is possible in the absence of cell division (Ray and Speck, 1973). The relevance of these observations to the nature of freeze injury will be discussed later.

FREEZE INJURY

Manifestation of Injury

The exact mechanisms by which freezing causes viability loss in microbial cells are not fully understood (Ray, 1983), but many studies have been conducted on the nature and site(s) of 'freeze-injury'. Some workers have identified gross changes, such as leakage of cell material, while others have probed directly for specific, macromolecular structural, functional or genetic changes in order to detect injury sites. From such studies, a host of characteristics have become recognisable which distinguish both lethally and non-lethally injured cells from healthy cells.

The ways by which freeze-injury has been demonstrated include: loss of viability; leakage of cellular materials; increased sensitivity to surfactants and other compounds; increased nutritional need; extended lag; and increased sensitivity to radiation (Ray and Speck, 1973; Macleod and Calcott, 1976).

Viability

Loss of viability is usually detected by the failure of cells to develop colonies on appropriate non-selective media under optimal growth conditions, at which stage the cells are regarded as dead even though such cells do not affect optical density or microscopic methods of estimating populations (Ray and Speck, 1973). In addition, some cells could fail to develop colonies in the presence of selective agents as a result of freeze injury; this is considered to result from damage to the protective cell barriers (Calcott, 1978).

Leakage

Some workers have reported a correlation between loss of viability and the quantity of cellular material leaked from frozen cells. Lindeberg and Lode (1963) found that death in frozen *E. coli* was proportional to the amount of nucleic acid lost through leakage. Similarly, Gritsavage (1965) observed a correlation between the concentrations of leaked protein-like substances and adenosine monophosphate compared with the extent of death in frozen *E. coli* cells. Calcott and Macleod (1975) demonstrated a release of UV-absorbing material which was inversely proportional to the survival of frozen cells of *E. coli*. Moss and Speck (1966), however, did not find any such correlation between loss of viability in frozen *E. coli* and leaked 260 nm absorbing materials. Other workers who have reported non-correlation between loss of viability and leaked materials in bacteria include Morichi (1969) and Kocka and Bretz (1969). It would appear that leakage is dependent on cooling rate, and that the quality of the leaked substances is of greater importance than the quantity in so far as the compounds leaked are essential to the viability of the cell (Moss and Speck, 1966).

Other materials that are known to have been leaked into the suspending medium of frozen cells include: biologically active peptides (Moss and Speck, 1966), cellular proteins (Ray and Speck, 1973), alkaline phosphatase and RNA (Morichi, 1969), DNA, glutamic dehydrogenase, isocitric dehydrogenase and amino acids including aspartate, threonine, serine, glutamate, glycine, alanine and histidine (Gabis, 1970; Swartz, 1971).

Increased sensitivity to surfactants and other compounds

The increased sensitivity of some Gram-negative bacteria to surface-active agents and other compounds has been demonstrated. Such substances, at the normal level of use, are tolerated by healthy cells, and this has facilitated the distinction between the structurally injured and the unharmed portion of cells in a frozen population (Ray and Speck, 1973). Non-lethally injured, frozen E. coli cells have been shown to fail to multiply and develop colonies in the presence of bile salts or deoxycholate on agar media (Bretz and Hartsell, 1959; Ray and Speck, 1972). Sensitivity to lysozyme arising from freezing and resulting in lysis in E. coli has also been reported (Kohn, 1960; Ray et al., 1972; Ray, 1983). In addition E. coli was reported to have developed sensitivity to actinomycin D, which normally does not penetrate the Gram-negative cell (Bretz and Kocka, 1967; Ray and Speck, 1972). These observations led to the suggestion that the barriers which normally protect cells become impaired by freezing, thus allowing the compounds to penetrate into the cells to bring about their observed inhibitory effects (Ray and Speck, 1973).

Nutritional needs

The increased nutritional needs of metabolically-injured, frozen cells (Straka and Stokes, 1959; Morichi, 1969) prevent them from forming colonies on minimal media. However, supplementing the medium with certain nutrients, including meat extract, tryptones, peptones or vitamin-free casein, has been reported to enhance the viability of E. coli, Pseudomonas, Shigella sonnei, Streptococcus lactis and Aerobacter aerogenes on agar plates (Arpai, 1962; Nakamura and Dowson, 1962; Moss and Speck, 1963; Postgate and Hunter, 1963). The failure to form colonies on minimal medium was attributed to the inability of metabolically-injured cells to synthesise all the necessary nutrients required for growth and multiplication. The ability of Salmonella anatum and E. coli to repair their structural injury from freezing, given suitable nutrients in liquid medium, has also been demonstrated by Ray et al. (1972) and Ray and Speck (1972). They showed that complete recovery was obtained in tripticase soy broth, while K_2HPO_4, pyruvate and ATP also enhanced repair.

Repair

A further characteristic of non-lethally freeze-injured cells is their ability to repair under favourable conditions (Ray and Speck, 1973).

This is seen when frozen cells, which have acquired a sensitivity to a selective agent, progressively lose this sensitivity on holding in a repair or 'resuscitation' medium. Thus, a rapid increase in counts on a selective medium is observed, with no simultaneous increase on a non-selective medium (Ray and Speck, 1973; Ray *et al.*, 1976).

The repair of metabolic injury in *E. coli* has been shown to be aided by the addition of low-molecular-weight peptides, but was independent of *de novo* synthesis of proteins, RNA, DNA or mucopeptides. Adenosine triphosphate (ATP) synthesis was, however, required for repair (Ray and Speck, 1973).

Ray and Speck (1972, 1973) showed that the repair of injury induced by freezing *E. coli* was most rapid in complex nutrient media. They advocated the use of trypticase soy broth as a pre-enrichment medium to facilitate repair before attempting to enumerate injured coliform bacteria on selective agar media.

Injury and repair in various *Salmonella* serotypes, after freezing and thawing, were studied by Janssen and Busta (1973). The importance of energy synthesis in repair mechanisms was demonstrated by the observation that 2,4-dinitrophenol, an uncoupler of oxidative phos-phorylation, inhibited repair of injury. Raccach and Juven (1976) studied the effect of suspending and plating media on the recovery of *S. gallinarum* following freezing and thawing. The reduction in survival of frozen and thawed cells in selenite–cystine and in tetrathionate broths again emphasised the necessity for resuscitation of injured cells before selective enumeration.

Specific Cellular Targets

Gross morphological alterations

Nei (1959) has observed that the freezing of *Bacillus megaterium* brought about two distinct general changes as detected under the light microscope. One was attributed to mechanical destruction of the cell wall and cell membrane, while the second was a cytochemical change as characterised by the staining reaction. Rapatz and Luyet (1963) have also reported cell shrinkage in *E. coli* when cooled slowly at about $1°C\,min^{-1}$. When cooled rapidly, however, at greater than $100°C\,min^{-1}$, the cells were found to be normal in size but had many intracellular cavities which were presumed to be occupied by ice crystals.

Cell membrane

Various studies have shown that membrane damage, with associated permeability impairment, resulted not only in the loss of cellular material through leakage, but in addition, rendered cells more vulnerable to penetration by injurious substances from the environment (Postgate and Hunter, 1963; Macleod *et al.* 1967). Gabis (1970) showed that large quantities of ether-soluble materials were found to be released when *E. coli* cells were frozen to −20°C in phosphate buffer. Since over 90% of the total bacterial lipid content is to be found in the cell membrane, it was considered that the lipid loss indicated some kind of alteration to the normal structure and function of the membrane. Souzu (1973) also demonstrated the increased extractability of membrane lipids in yeast following freezing and thawing, and it was suggested that many covalent bonds were probably disrupted, thus changing the lipids from bound to unbound forms.

Membrane damage has been demonstrated by: increased salt sensitivity (Lee *et al.*, 1977); increased sensitivity to bile salts and deoxycholate (Ray and Speck, 1972); increased sensitivity to lysozyme (Kohn and Szybalski, 1959); increased sensitivity to EDTA (Ray *et al.*, 1972); release of periplasmic enzymes and alkaline phosphatase (Morichi, 1969); release of cyclic phosphodiesterase and β-galactosidase (Calcott and Macleod, 1975); release of malate, glucose-6-phosphate and glutamate dehydrogenases (Souzu, 1980).

Some workers (Smittle, 1973; Smittle *et al.*, 1974) have observed that the presence of a high concentration of unsaturated fatty acids and a low concentration of saturated fatty acids in the membrane enhances cryosurvival in *Lactobacillus bulgaricus*. It was suggested that this condition provides membrane fluidity and elasticity which reduces freezing damage to the membrane (Ray and Speck, 1973). Evidence consistent with this view is found in the reports of Steim *et al.* (1969) and Haest *et al.* (1972).

Steim *et al.* (1969) demonstrated thermal phase transition properties for both membranes and protein-free lipids of *Mycoplasma laidlawii* as detected by differential scanning calorimetry. The phase transition temperatures, which were found to be lowered by unsaturation in fatty acids, were found to be the same for both membranes and lipids. They suggested that the lipids were in a bilayer conformation in which the hydrocarbon chains associated with each other rather than with proteins, and that for proper transport process functioning, it was essential that the hydrocarbons were in a state of fluidity.

The above reasoning was further reinforced by the report of Haest *et al.* (1972), in which it was suggested that it was essential for the paraffin core to exist in the proper liquid condition for the membrane to function correctly, and that the quantity and quality of unsaturated fatty acids regulate membrane fluidity. In addition, it is thought that the cooling of a bacterial membrane below a certain critical temperature (transition temperature) might bring about the solidification of the paraffin core of the membrane, which could produce discontinuities in the membrane packing with the permitted release of leaked substances (Ray and Speck, 1973); furthermore, the greater the unsaturation, the lower the transition temperature (Steim *et al.*, 1969; Ellar, 1978).

Evidence that lower growth temperatures can bring about increased unsaturated fatty acid composition in the membrane has been shown by Farrell and Rose (1968) and Herbert and Bhakoo (1979). It would, therefore, seem that lower growth temperatures can bring about, among other things, a change in the quantity and quality of the unsaturated fatty acids in the membrane, with the consequent lowering of the transition temperature, increased fluidity and elasticity, and increased adaptability to cope with the injurious effects of freezing.

It has recently been shown (Souzu, 1980) for *E. coli* that the type of membrane damage is freezing-rate dependent. It was observed that slow freeze–thawing $(3\text{--}10^\circ\text{C min}^{-1})$ caused damage mainly to the outer membrane of *E. coli,* while rapid freeze–thawing $(200^\circ\text{C min}^{-1}$ cooling) caused damage to both the cytoplasmic membrane and the outer membrane.

Cell wall

The possession of the lipopolysaccharide layer (LPS) in some Gram-negative bacteria is considered to be a protection against several surface-active and otherwise injurious compounds, as well as preventing the exit of various molecules. However, damage of this outer membrane (Costerton *et al.*, 1974) by freezing has been demonstrated with the use of actinomycin D (Leive, 1965; Singh *et al.*, 1972), bile salts and deoxycholate (Ray *et al.*, 1972), lysozyme and trypsin (Kohn, 1960) and bacteriophages (Kempler and Ray, 1978; Ray, 1983).

Nucleic acids

The release of RNA, but not DNA, from frozen cells has been reported by Gabis (1970) and Morichi (1969), and of UV-absorbing material by Calcott and Macleod (1975). It was suggested by Morichi (1969) that freezing might have interfered with the binding of ribosomes to the membrane and activated the latent RNAse present in the ribosomes, thus allowing the degradation of the disrupted ribosomes. Single-strand breaks in DNA because of freezing have been reported by Swartz (1971), Takano *et al.* (1973) and Alur and Grecz (1975), while Macleod and Calcott (1976) have reported the detection of double-strand breaks in DNA following freezing and thawing. However, others (Ashwood-Smith *et al.*, 1972) were unable to detect any DNA breaks in freeze-injured cells.

Proteins

The detection of proteins in the freezing menstruum after the freeze–thawing of cells is well documented, and Gabis (1970) has, in addition, observed that large quantities of free amino acids are present in the freezing menstruum of frozen *E. coli* cells. As a result, he proposed that the freezing process might have activated some latent proteolytic enzymes, thereby bringing about an increased breakdown of proteins. He further suggested that the presence of large amounts of basic amino acids, along with glycine, alanine and leucine, might indicate that the ribosomal proteins (known to be rich in those amino acids) had been hydrolysed in the frozen cells. Generally, freezing and thawing are said to cause a partial unfolding of the helical structure of fibrous proteins without any conformational change in globular protein, because of the increased conformational stability of the latter (Hanafusa, 1967). Reay (1933) and Moran (1935) have also ascribed death of frozen micro-organisms to protein denaturation, after observing that increased mortality, at −2°C as against −20°C, was accompanied by the precipitation of the coagulable proteins of the organsims.

Alterations in Cell Functions

Enzyme activity

As reported above, Gabis (1970) has suggested that freezing could activate some otherwise latent proteolytic enzymes. The same author, however, did not find any significant changes in the activity of isocitric- and glutamic-dehydrogenase enzymes of the same frozen cells. The loss of some enzyme activity in lactic acid bacteria has been reported, involving both acid producing ability and proteolytic activity, during frozen storage at −20°C (Cowman and Speck, 1965). However, upon using −196°C for freezing and storage, it was observed that both acid producing ability and proteolytic activity were retained. The same authors (Cowman and Speck, 1969) observed that the membrane-bound proteinase enzyme of *S. lactis* needed to remain in monomer–dimer equilibrium for normal action. It was found that freezing and frozen storage of the enzyme at −20°C for 3 days brought about polymerisation into a high-molecular-weight form, with concomitant loss of enzymatic activity.

Pathogenicity

The effects of freezing and thawing have been shown not to cause any loss in pathogenicity of *Salmonella gallinarium* (Sorrells *et al.*, 1970). Hollander and Nell (1954) on the other hand, previously reported that *Treponema pallidum*, when frozen in saline or serum to −78°C, required a prolonged incubation period when injected into rabbits after thawing. Repeated freeze–thawing for four times was found to result in complete loss of infectivity, and the use of glycerol as the freezing medium did not improve infectivity after freezing and storage for 2 months at −78°C.

Respiration

The freezing of bacterial cells has been shown to affect their respiratory activity in several ways. O'Hara (1954) reported that freezing and thawing increased the respiratory activity of *E. coli* cells, but, while repeated freeze–thawing was found to progressively increase the respiratory activity for the first several cycles, it decreased later until totally lost at the 40th cycle. On the contrary, Sato (1954) reported a decreased oxygen uptake arising from the freezing and thawing of *Bacillus megaterium*. Postgate and Hunter (1963) have observed, however, that the temperature of freezing might affect the respiratory activity of cells. They found that cells of *Aerobacter aerogenes* frozen rapidly at −196°C showed a much lower O_2 uptake: CO_2 production ratio than those frozen at −4°C.

Oxidative phosphorylation

Not much information has been provided, as yet, on the effect of freeze–thawing on the processes of oxidative phosphorylation. Nevertheless, Aithal *et al.* (1971) demonstrated that electron transport particles from *Mycobacterium phlei,* frozen rapidly to −75°C for 10 min then thawed slowly at room temperature, exhibited a reduced P:O ratio; phosphorylation was found to decrease by 50% without any significant decrease in oxidation. Freezing in liquid nitrogen, on the other hand, brought about decreases in both oxidation and phosphorylation. It was further observed that ATP and Mg^{++}, plus soluble coupling factors and cryoprotective agents, protected against the damaging effects of freezing. Hence they suggested that the freezing-induced reduction in phosphorylation activity occurred probably through some alteration in the organisation of the electron transport particles within the membrane structure. Lee, Calcott and Macleod (1977) have demonstrated that sensitivity to salt during slow freezing and thawing is related to the presence of cytochromes in the cytoplasmic membrane. Cells, such as *Streptococcus,* which lacked cytochromes were relatively insensitive to salt in the freezing menstruum.

Mechanisms of Freeze Damage

Various theories have been proposed to explain the mechanisms of death of bacterial cells caused by freezing (Calcott, 1978). One of the earliest proposals attributed death to the mechanical/structural damage of cells caused by the crushing and spearing action of ice crystals (Keith, 1913; Hillard and Davis, 1918; Weiser and Osterud, 1945). Others believed that lethality resulted from cell dehydration effects during freezing (Harrison, 1956). Unlike the earlier group, which considered that ice was mainly extracellular, many authors have asserted that lethality is entirely caused by intracellular ice formation (Sato, 1954; Mazur, 1966; Nei *et al.,* 1967, 1969). This has been verified by electron microscopy where rapidly frozen cells of *E. coli* (Nei, 1973) and yeast (Bank and Mazur, 1973) have been shown to contain cavities presumed to have been occupied by ice crystals.

The above views were consolidated into the 'two factor hypothesis' of Mazur (1966, 1970), i.e. one factor causes lethality at low cooling rates and the other at higher rates. Death results from the effects of both factors, and the effects are minimal at the optimum cooling rate for survival for each cell type. It has been convenient to regard these factors as: solute concentration affecting cells at slow freezing rates and intracellular ice causing damage in rapidly frozen cells. A possible scenario is given below.

When a system is frozen, the temperature drop initially results in supercooling of the medium followed by ice crystal formation. The intracellular water, however, remains unfrozen and supercooled until the temperature drops down to $-10°C$ to $-15°C$, above which only extracellular ice crystallisation is expected to occur. The extracellular ice formation lowers the vapour pressure of the medium and the cell reacts to re-establish osmotic equilibrium either by freezing cellular water inside the cell, to reduce its 'availability', or by losing it and freezing extracellularly. The permeability of cell membranes to water and the surface–volume ratios of cells are, therefore, important factors. At low cooling rates/high membrane permeability/high surface–volume ratios ice forms extracellularly and cells dehydrate and shrink. Under these conditions, lethal effects result from high solute concentrations inside the cells. At high cooling rates/low membrane permeability/low surface–volume ratios external ice grows through the water-filled pores of the membranes to nucleate the internal, supercooled cellular water which freezes intracellularly. At ultra-rapid freezing rates, these intracellular ice crystals are very small and hence less mechanically disruptive. Slow thawing, however, permits them to coalesce and grow with more lethal consequences to the cell.

Studies of freeze damage to the cell envelope of *E. coli* (Souzu, 1980), while supporting the generalised views, also pointed out that the lysis of cell membranes should not simply be attributed to ice crystal formation. The cell envelope of Gram-negative bacteria has at least three, distinct, macromolecular components: an inner cytoplasmic membrane with phospholipids and proteins; a peptidoglycan layer external to the inner membrane, and an outer membrane containing lipopolysaccharide (LPS) in addition to phospholipids and protein (Costerton *et al.*, 1974). Membrane function is greatly influenced by the fluidity, and thus the physical state, of the membrane components. The bipolar relationships of membrane proteins and lipids are changed at sub-zero, phase-transition temperatures which reflect the melting points of the membrane lipids (Bayer *et al.*, 1977). Souzu demonstrated that freeze–thawing causes the release of cell membrane components from both inner and outer membranes. However, slow freeze–thawing, which causes greater loss of cell viability, results in more release of LPS and outer membrane components and loss of a periplasmic dehydrogenase, but negligible release of cytoplasmic marker enzymes. Rapid freeze–thawing ($200°C$ min^{-1}), on the other hand, results in less release of outer membrane material, but causes more cytoplasmic

membrane damage. The former effect was attributed to the fragmenta-
tion of the outer membranes at their phase-transition temperatures, a
change that is avoided in the latter case by the speed at which
temperatures fall through the critical transition values before all the
lipids crystallised. Damage, in the case of rapidly frozen cells, was
thought to ensue from the movement of cellular water through the
crystallised inner membrane to equilibrate osmotic pressure differences
as extracellular ice formed. Souzu pointed out that the 'rapid' cooling
rate of $200°C\,min^{-1}$ was marginally below the cooling velocity
considered minimal for intracellular ice formation by Mazur (1965).
The recent studies of Ray (1983), which addressed the controversy
concerning the site and mechanisms of cryoprotection on microbial
cells, also demonstrated outer membrane damage in *E. coli* as a
function of freezing rate; lysozyme lysis and the ability to adsorb
LPS-specific bacteriophages were used as indicators of cell surface
integrity. Cryoprotectants which penetrated the cells provided more
effective protection against freeze damage than those which were
non-permeating and remained on the cell surfaces. The superiority of
permeating cryoprotectants supports Ray's view that viability loss
caused by freezing is not only caused by damage to the cytoplasmic
membrane, but may also be caused by cumulative effects of damage to
different structural and functional macromolecules (Ray and Speck,
1973; Ray *et al.*, 1976).

Despite extensive study and thoughtful review, the basic mechanisms
of freeze lethality in micro-organisms have not yet been clearly
elucidated. Many of the original questions of Mazur (1966) remain
essentially unanswered: What are the conditions for intracellular ice
formation in various cells? What are the actual effects of intracellular
freezing on membranes and on cellular macromolecules? What are the
mechanisms of cryoprotection? How do concentrated solutes cause loss
of cell viability?

REFERENCES

Aithal, H. N., Kalra, V. K. and Brodie, A. F. (1971). *Biochem. Biophys. Res.
    Com.*, **43**, 550.
Alur, M. D., and Grecz, N. (1975). *Biochem. Biophys. Res. Com.*, **62**, 308.
Arpai, J. (1962). *Appl. Microbiol.*, **10**, 297.
Ashwood-Smith, M. J., Trevino, J. and Warby, C. (1972). *Cryobiology*, **9**,
    141.
Bank, H. and Mazur, P. (1973). *J. Cell. Biol.*, **57**, 729.

Bayer, M. E., Dolack, M. and Houser, E. (1977). *J. Bacteriol.*, **129**, 1563.
Bretz, H. W. and Hartsell, S. E. (1959). *Food Res.*, **24**, 369.
Bretz, H. W. and Kocka, F. E. (1967). *Can. J. Microbiol.*, **13**, 914.
Calcott, P. H. (1977). In: *CRC Handbook of Nutrition and Food* (ed. M. Richcigl), CRC Press, Cleveland.
Calcott, P. H. (1978). In: *Freezing and Thawing Microbes, Patterns of Progress*, Meadowfield Press Ltd, Shildon, Co. Durham.
Calcott, P. H. and Macleod, R. A. (1974*a*). *Can. J. Microbiol.*, **20**, 671.
Calcott, P. H. and Macleod, R. A. (1974*b*). *Can. J. Microbiol.*, **20**, 683.
Calcott, P. H. and Macleod, R. A. (1975). *Can. J. Microbiol.*, **21**, 1724.
Costerton, J. W., Ingram, J. M. and Cheng, K. J. (1974). *Bacteriol. Rev.*, **38**, 87.
Cowman, R. A. and Speck, M. L. (1965). *J. Dairy Res.*, **48**, 1531.
Cowman, R. A. and Speck, M. L. (1969). *Cryobiology*, **5**, 291.
Davies, J. E. (1970). In: *The Frozen Cell* (Eds G. E. W. Wolstenholme and M. O'Connor), Churchill, London.
Ellar, J. D. (1978). In: *Companion to Microbiology* (Eds A. T. Bull and P. M. Meadow), Longman, London.
Elliott, R. P. and Michener, H. D. (1965). *Psychrophilic Micro-organisms in Foods*, Technical Bulletin No 1320, USDA.
Farrell, J. and Rose, A. H. (1968). *J. Gen. Micribiol.*, **50**, 429.
Gabis, D. A. (1970). Studies on the effects of low temperature storage on the metabolism of *Escherichia coli*. PhD Thesis, North Carolina State University.
Gilliland, S. E. and Speck, M. C. (1974). *Appl. Microbiol.*, **27**, 793.
Gould, G. W. and Dring, G. J. (1975). *Nature*, **258**, 402.
Gritsavage, E. F. (1965). The cytochemical effect of freezing on *Escherichia coli*. PhD Thesis, Michigan State University.
Haest, C. W. M., DeGier, J., Van Es, G. A., Verkleu, A. J. and Van Deenan, L. L. M. (1972).*Biochim. Biophys. Acta*, **288**, 43.
Hanafusa, N. (1967). In: *Cellular Injury and Resistance in Freezing Organisms* (Ed. E. Asahina), Hokkaido University, Sapporo, Japan.
Harrison, A. P. (1956). *Antonie Van Leeuwenhoek*, **22**, 407.
Herbert, R. A., and Bhakoo, M. (1979). In: *Cold Tolerant Microbes in Spoilage and the Environment* (Eds. A. D. Russell and R. Fuller), Academic Press, London.
Hillard, C. M. and Davis, M. A. (1918). *J. Bacteriol.*, **3**, 423.
Hollander, D. H. and Nell, E. E. (1954). *Appl. Microbiol.*, **2**, 164.
Ingram, M. and Mackey, B. M. (1976). In: *Inhibition and Inactivation of Vegetative Microbes,* (Eds F. A. Skinner and W. B. Hugo), Academic Press, London.
Janssen, D. W. and Busta, F. F. (1973). *J. Milk Food Technol.*, **36**, 520.
Jay, J. M. (1978). *Modern Food Microbiology*, Van Nostrand Co., New York.
Johannsen, E. (1972). *J. Appl. Bacteriol.*, **35**, 415.
Keith, S. C. (1913). *Science*, **37**, 877.
Kempler, G. and Ray, B. (1978). *Cryobiology*, **15**, 578.
Kocka, F. E. and Bretz, H. W. (1969). *Antonie Van Leeuwenhoek*, **35**, 65.
Kohn, A. (1960). *J. Bacteriol.*, **79**, 697.

Kohn, A. and Szybalski, W. (1959). *Bact. Proc.*, 126.
Larkin, J. M. and Stokes, J. L. (1968). *Can. J. Microbiol.*, **14**, 97.
Lee, S. K., Calcott, P. H. and Macleod, R. A. (1977). *Can. J. Microbiol.*, **23**, 413.
Leive, L. (1965). *Biochem. Biophys. Res. Com.*, **21**, 290.
Lindeberg, G., and Lode, A. (1963). *Can. J. Microbiol.*, **9**, 523.
Macleod, R. A. and Calcott, P. H. (1976). In: *Survival of Vegetative Bacteria* (Eds T. R. G. Gray and J. R. Postgate), Cambridge University Press, Cambridge.
Macleod, R. A., Kuo, S. C., and Gelinas, R. (1967). *J. Bacteriol.*, **93**, 961.
Mazur, P. (1965). *Ann. N. Y. Acad. Sci.*, **125**, 658.
Mazur, P. (1966). In: *Cryobiology* (Ed. H. T. Meryman), Academic Press, New York.
Mazur, P. (1970). *Science*, **168**, 939.
Meryman, H. T. (1966). In: *Cryobiology* (Ed. H. T. Meryman), Academic Press, New York.
Meryman, H. T. (1974). *Ann. Rev. Biophys. Bioeng.* **3**, 341.
Moran, T. (1935). *J. Soc. Chem. Ind.*, **54**, 149.
Morichi, T. (1969). In: *Freezing and Drying of Micro-organisms* (Ed. T. Nei), University of Tokyo Press, Tokyo.
Moss, C. W. and Speck, M. L. (1963). *Appl. Microbiol.*, **11**, 326.
Moss, C. W. and Speck, M. L. (1966). *J. Bacteriol.*, **91**, 1098.
Nakamura, M. and Dowson, D. A. (1962). *Appl. Microbiol.*, **10**, 40.
Nei, T. (1959). In: *Recent Researches in Freezing and Drying* (Eds A. S. Parkes and A. U. Smith), Blackwell, Oxford.
Nei, T. (1973). *Cryobiology*, **10**, 403.
Nei, T., Araki, T. and Matsusaka, T. (1967). In: *Cellular Injury and Resistance in Freezing Organisms* (Ed. A. Asahina), Hokkaido University, Sapporo, Japan.
Nei, T., Araki, T. and Matsusaka, T. (1969). In: *Freezing and Drying of Micro-organisms,* (Ed. T. Nei), University Park Press, Baltimore.
Obafemi, A. (1983). The survival of *Salmonella typhimurium* in processed frozen poultry. PhD Thesis, University of Reading, England.
O'Hara, Y. (1954). *Biol. Sci.*, **12**, 37.
Olson, J. C. and Nottingham, P. M. (1980). In: *Microbial Ecology of Foods, Vol. I.,* ICMSF, Academic Press, New York.
Packer, E. L., Ingraham, J. L. and Scher, S. (1965). *J. Bacteriol.*, **89**, 718.
Partmann, W. (1975). In: *Water Relations of Foods* (Ed. R. B. Duckworth), Academic Press, New York.
Postgate, J. R. and Hunter, J. R. (1963). *J. Appl. Bacteriol.* **26**, 405.
Raccach, M. and Juven, B. J. (1976). *J. Food Technol.*, **11**, 221.
Rapatz, G. and Luyet, B. (1963). *Abs. Biophys. Soc.*, New York.
Ray, B. (1983). *J. Food Protection,* **46**, 864.
Ray, B. and Speck, M. L. (1972). *Appl. Microbiol.*, **24**, 585.
Ray, B. and Speck, M. L. (1973). *CRC Critical Reviews, Clinical Laboratory Science,* **4**, 161.
Ray, B., Janssen, D. W. and Busta, F. I. (1972). *Appl. Microbiol.*, **23**, 803.
Ray, B., Speck, M. L. and Dobrogosz, W. J. (1976). *Cryobiology*, **13**, 153.

Reay, G. A. (1933). *J. Soc. Chem. Ind.*, **52,** 265.

Sato, T. (1954). *Biol. Sci.*, **12,** 39.

Singh, A. P., Cheng, K. J., Costerton, J. W., Idziak, E. S. and Ingram, J. M. (1972). *Can. J. Microbiol.*, **18,** 909.

Smittle, R. B. (1973). Relationships of growth medium to death of *Lactobacillus bulgaricus* in liquid nitrogen. PhD Thesis, North Carolina State University.

Smittle, R. B., Gilliland, S. E., Speck, M. L. and Walter, W. M. (1974). *Appl. Microbiol.*, **27,** 738.

Sorrells, K. M., Speck, M. L. and Warren, J. W. (1970). *Appl. Microbiol.*, **19,** 39.

Souzu, H. (1973). *Cryobiology*, **10,** 427.

Souzu, H. (1980). *Biochim. Biophys. Acta*, **603,** 13.

Straka, R. P., and Stokes, J. L. (1959). *J. Bacteriol.*, **78,** 181.

Steim, J. M., Tourtelotte, M. E., Reinert, J. C., McElhaney, R. N. and Rader, R. L. (1969). *Proc. Nat. Acad. Sci.*, **63,** 104.

Swartz, H. M. (1971). *Cryobiology*, **8,** 255.

Takano, M., Sinskey, A. J. and Baraldi, D. (1973). *IIF–IIR Commission C1—Sapporo*, **5,** 61.

Takano, M., Simbol, A. B., Yasin, M. and Shibasaki, I. (1979). *J. Food Sci.*, **44,** 112.

Toyokawa, K. and Hillander, D. H. (1956). *Proc. Soc. Exptl. Biol. Med.*, **92,** 499.

Trakulchang, S. P., and Kraft, A. A. (1977). *J. Food Sci.*, **42,** 518.

van den Berg, L. (1968). In: *Low Temperature Biology of Foodstuffs* (Eds J. Hawthorn and E. J. Rolfe), Pergammon Press, London.

Weiser, R. S. and Osterud, C. M. (1945). *J. Bacteriol.*, **50,** 413.

Chapter 4

Microbiology of Frozen Meat and Meat Products

P. D. Lowry and C. O. Gill

Meat Industry Research Institute of New Zealand, Hamilton, New Zealand

INTRODUCTION

Meat is a highly perishable food commodity, which, unless appropriately modified or stored, will rapidly spoil from the growth of micro-organisms. To prolong the shelf-life of meat it is desirable not only to retard general microbial growth, but to adjust conditions to retard specifically growth of organisms of high spoilage potential. These objectives can be achieved either by direct alteration of the physico-chemical properties of the meat, or by manipulation of the storage conditions. Many traditional techniques of meat preservation, e.g. drying or addition of curing salts, are based on the former method; they inhibit microbial growth principally by reducing water activity either through water removal or solute addition. Unfortunately, these techniques also markedly alter the organoleptic properties of meats, so that they bear little similarity to the fresh product. The storage of meats at chilled temperatures overcomes these difficulties, and the retardation of spoilage occurs with only minimal effects on the sensory attributes of the chilled meat. Nevertheless, the temperature tolerances of the organisms that usually make up the spoilage microfloras of fresh meats are such that even at chilled temperatures, spoilage levels will be attained within two or three weeks. Therefore if microbial spoilage is to be entirely precluded, storage temperatures must be reduced to below the minimum for microbial growth, i.e. below the temperature at which freezing of tissues will occur.

Micro-organisms differ in their responses to the freezing of their environment. Some survive virtually unharmed; some resist freezing but are susceptible to damage during frozen storage or thawing; others are sensitive to frozen storage and thawing under only some conditions; and others are inactivated by freezing under nearly all conditions (ICMSF, 1980*a*). These effects are not simply responses to temperatures too low to permit growth; rather, they reflect the physical changes to the microbial environment that occur when freezing removes water from the tissues. The effects of such changes have been studied extensively for micro-organisms suspended in laboratory media. These studies have elucidated the mechanisms by which micro-organisms may be damaged and destroyed during freezing, and have shown that microbial survival of freezing is, quantitatively, highly dependent on the freezing regime employed and the composition of the suspending medium. As there are few detailed data on how freezing affects the microfloras on meat, data from studies on frozen meats must necessarily be interpreted using knowledge acquired from laboratory systems employing artificial media. In so doing, it is essential that data appropriate to meat systems are distinguished from data relating to circumstances unlikely to occur during the processing and storage of meats. This chapter examines the conditions to which micro-organisms are exposed upon the freezing of meat, so that the practical consequences of the freezing of meats and meat products for meat microfloras can be described.

'Meat' can be defined as the flesh of any animal that is used for food (Lawrie, 1974). Although there are many basic similarities in the microfloras and biochemical compositions of the various tissues encompassed by the above definition, a comprehensive consideration of the effects of freezing on microfloras from all meat tissues is beyond the scope of this chapter. Therefore, discussion will centre on the effects of freezing on the microfloras of muscle tissue from the most common domestic food animals: cattle, pigs and sheep. The specific cases of poultry meat and mammalian organ tissues are also considered, but in separate sections. For other meats, the literature either is inadequate or does not distinguish any peculiar qualities with respect to freezing. Therefore, such meats are not given consideration.

Similar, somewhat arbitrary limitations are required for effective discussion of meat products. Broadly defined, these foods include those consisting almost entirely of meat, through those with admixtures of various levels of non-meat protein, fat and/or carbohydrate, to those in which meat is only a minor component. Only the first group will be considered, as the inclusion of large proportions of non-meat material introduces physical, chemical and microbiological elements that are very different from those obtained with whole meat. The meat products discussed are, therefore, essentially 'processed meats' (Hannan, 1975); that is, meat containing various materials (added for preservation or flavour modification) which may extensively change the physical properties of the meat, but do not grossly alter its basic chemical composition.

THE EFFECT OF FREEZING ON MICRO-ORGANISMS

The mechanisms by which freezing may damage micro-organisms have been reviewed by Mazur (1966), Hagen (1971) and Ray and Speck (1973). These reviews comprehensively describe the freezing-related phenomena that affect micro-organisms, and this discussion need only summarise the major conclusions.

In principle, five factors could cause cell damage during freezing: low temperature; extracellular ice formation; intracellular ice formation; concentration of extracellular solutes as the extracellular water is progressively frozen out as pure ice; and concentration of intracellular solutes as cell water migrates to the freezing and concentrating extracellular medium.

Low temperature and extracellular ice formation *per se* are not causes of cell damage. Most micro-organisms can remain viable after dehydration, so intracellular solute concentration likewise is not a major cause of freezing injury. Rather, cell damage on freezing is due to intracellular ice formation and/or concentration of the extracellular solutes.

Intracellular freezing will only occur under a limited range of circumstances, for the cell cytoplasm will invariably supercool to temperatures approaching $-10°C$. The osmotic differences between the freezing external medium and supercooled cytoplasm will result in the cell tending to lose water to the environment, the absolute rate of which depends on both the difference in water vapour pressure between the cytoplasm and the medium, and the water permeability of the cell. The rate of dehydration is also highly dependent on cell size, smaller cells losing a greater proportion of their water more rapidly than larger cells. If cooling is sufficiently slow for a particular cell to lose water at a rate that will maintain the cytoplasm close to osmotic balance with the environment, the intracellular solute concentration will increase proportionally with the temperature decrease and intracellular freezing of water will never occur. Hence, intracellular freezing only occurs if the rate of cooling is so rapid that the rate of water loss from the cell is too slow to maintain near equilibrium conditions, and the limits to cytoplasmic supercooling are exceeded. In general, cooling rates of $1°C \, min^{-1}$ or less will be too slow for intracellular ice formation to occur in any micro-organism. For large cells, such as yeasts, cooling rates in excess of $10°C \, min^{-1}$ can produce intracellular ice formation, but for the smaller cells of bacteria, cooling rates probably must exceed $100°C \, min^{-1}$ before intracellular ice formation will occur.

By contrast, any freezing-out of pure ice from a medium will expose cells in the remaining liquid phase to elevated solute concentrations, and as the ice and residual solution will be in equilibrium with equivalent vapour pressures, the concentration of the solute will depend on the vapour pressure of ice, which, in turn, is determined solely by the temperature. The concentration of extracellular solutes *per se* does not cause cell injury, and indeed some solutes give increased protection against freezing injury as their concentrations increase. Cell damage appears, therefore, to be related to the concentration of specific solutes, and the effective solute species may differ among organisms. For example, the presence of NaCl in a medium will increase freezing injury of the bacterium *Escherichia coli*, but will protect the yeast *Saccharomyces cerevisiae*.

Although the basic causes of cell injury and death during freezing are known, quantitative effects cannot generally be predicted, because many factors can significantly influence survival rates. Relevant factors include rates of cooling and warming during freezing and thawing, the chemical composition of the microbial environment, the physiological state of the micro-organisms, and inherent characteristics of individual species that predispose them to susceptibility or resistance to freezing injury. Even with model systems, all these factors cannot conveniently be monitored or controlled simultaneously. Consequently, different workers using similar model systems have arrived at various, sometimes contradictory, conclusions as to the lethal effects of freezing on specific micro-organisms. Consistency of detail cannot, therefore, be expected of data obtained for microfloras in the more complex and variable systems presented by meat and meat products undergoing freezing and thawing. However, by consideration of commercial circumstances, it should be possible to define the conditions to which micro-organisms may be exposed during the freezing of meat and, by application of data from models relevant to the behaviour of typical meat micro-organisms, to derive a generalised view of the effects of freezing on meat microfloras.

MEAT COMPOSITION

As microbial survival of freezing is determined, in part, by the composition of the medium in which freezing occurs, the environments to which micro-organisms on meat may be exposed must be considered. Meat comprises various proportions of muscle, adipose and connective tissues. Adipose and connective tissues are very different in composition to muscle tissue, but microbial growth on muscle tissue is what usually determines the shelf-life of meat stored under conditions that allow microbial proliferation.

Muscle-tissue composition can vary even between muscles from the same carcass, but the relative proportions of the bulk components fall within a narrow range. Water makes up 70–79% of the muscle-tissue weight (Fennema and Powrie, 1964); protein contributes about 18%, and lipid approximately 3% (Lawrie, 1974). The remaining fraction, approximately 3·5%, is composed of inorganic materials and variable proportions of low-molecular-weight, soluble organic compounds (Table I). These latter, quantitatively minor components, are major determinants of the environment provided to micro-organisms by meat.

The level of inorganic and nitrogenous soluble components do not vary greatly between living and post-rigor muscle, although biochemically important changes occur in nucleotide composition, and amino acid concentrations tend to rise with extended storage. However, the concentrations of substances derived from glycogen via glycolysis change significantly. Post-mortem changes in muscle leading to rigor development involve degradation of stored glycogen, the main end-product being lactic acid. Concentrations of glycolytic intermediates fluctuate during rigor development, but finally fall to low levels. Usually enough lactic acid accumulates for the pH of muscle to fall from the neutral values of living tissue to an ultimate value between 5·5 and 5·8 (Bendall, 1973). If, however, exercise or stress before slaughter

TABLE I
Composition of the low-molecular-weight, soluble
materials of post-rigor beef *sternomandibularis* muscle. (Data from Winger and Pope (1981) unless
otherwise indicated.)

Substance	Concentration	
	$mmol\,kg^{-1}$ tissue	$mmol\,kg^{-1}$ water
K	102·3	134·2
Na	25·0	32·8
Mg	9·6	12·6
Ca	1·0	1·3
Fe	1·1	1·4
Zn	0·8	1·1
P	60·0	78·7
S	18·1	23·7
Cl	9·1	12·0
Carnosine	9·0	11·8
Anserine	1·2	1·6
Creatine	28·0	36·8
Amino Acids	9·9	13·0
Nucleotides	5·1	6·7
Lactic Acid	105	138
Glucose[a]	0·6	0·8
Glucose-6-phosphate[a]	0·7	0·9

[a] Calculated from data in Gill (1985).

caused some of the muscle's glycogen reserve to be depleted,
decreased amounts of lactic acid will be formed, and other minor
derivatives of glycogen, such as glucose, may be absent. Meat in this
condition will have an ultimate pH in excess of 6·0, and, for beef, a
characteristic dark, firm, dry (DFD) appearance (Wirth, 1976).

A second atypical, stress-induced condition of meat, referred to
from its appearance as pale, soft, exudative (PSE), is associated mainly
with pork, although a mild form has been reported for beef (Fischer
and Hamm, 1978). The condition is genetically determined, and is
caused by protein denaturation consequent upon very rapid rigor
development; this causes the pH to reach low values while the tissue is
still warm. Characteristically, the meat is pale, has a low water-holding
capacity, and a pH at or below 5·5. The PSE condition has little effect
on the microflora of fresh meat, as its soluble-component composition
is similar to that of normal muscle, but the poor water-holding capacity

results in excessive drip, and also high solids losses during the cooking of processed meat (Cassens *et al.*, 1975).

Adipose tissue is composed largely of triglycerides, but small amounts of low-molecular-weight, soluble components are provided at fat surfaces by serum from damaged blood vessels. In contrast to muscle-tissue surfaces, the lactic acid concentration is low, so the pH of fat surfaces remains near neutrality; glycolytic intermediates are absent, although glucose and amino acids are present (Gill and Newton, 1980*a*).

From the viewpoint of microbial development, surfaces covered by connective tissue will provide the same soluble components as the underlying muscle or adipose tissue.

THE PHYSICAL ENVIRONMENT PROVIDED BY MEAT

The physical environment provided for micro-organisms by meat is also variable. For micro-organisms within a comminuted meat mass, the environment will be described by the bulk composition of muscle tissue. However, conditions may be very different for micro-organisms present at a meat surface. After flaying, all meat surfaces tend to dry as water evaporates from the warm carcass (Nottingham, 1982). The resultant surface desiccation precludes microbial growth until the tissues rehydrate as a result of water movement from the underlying bulk, growth commencing once there is adequate diffusion to compensate for the reduced evaporative loss from the cool carcass. Generally, surface desiccation of carcasses held at moderate ambient temperatures will delay microbial growth for at least 24 h, provided that the carcasses are sufficiently well spaced to allow air circulation (Scott and Vickery, 1939). If carcasses are exposed to chilling temperatures and forced-draught air circulation, the period of surface desiccation will be greatly prolonged.

In the case of fat surfaces, rehydration by water movement from the underlying tissue will be extremely slow compared with that at lean meat surfaces. Fat surfaces will, therefore, tend to remain dry unless storage conditions allow condensation of water vapour or some other process of direct wetting.

Micro-organisms on dry surfaces will have been exposed to solute concentration effects at ambient temperatures, and will be in the

116   *P. D. Lowry and C. O. Gill*

stationary phase when exposed to freezing temperatures. Freezing of the bulk substrate has no significance for organisms in this state apart from the temperature decrease that is involved. All intermediate stages of meat-surface desiccation prior to freezing are obviously possible, and this factor may be a significant source of variation in the response of the microflora to freezing.

THE FREEZING TEMPERATURE OF MEAT

The temperature at which freezing of the water in meat commences will be determined by the concentration of solutes present. The concentration of solutes in pre- and post-rigor muscle tissue is very different, with the osmotic pressures of the tissues being about 300 mOs and 500 mOs respectively (Winger and Pope, 1981). The freezing of pre-rigor muscle is not a usual or desirable commercial practice because of the resultant tissue toughening (Locker *et al.*, 1975), so pre-rigor freezing is not, in practice, an important cause of variation in the temperature of freezing onset. However, lactate concentrations will be significantly reduced in DFD muscle, which may, therefore, begin to freeze at a somewhat higher temperature than normal muscle. In addition, solute concentrations in liquid films on fat surfaces may be very much lower than in muscle, so that at fat surfaces extensive freezing may commence at temperatures little below 0°C.

TABLE II

Relationship between sub-freezing temperature, per cent frozen water and water activity in meat samples containing ice. (Data from Reidel (1957) and Leistner *et al.* (1981).)

Temperature (°C)	Percent frozen water in lean meat (74·5% H_2O)	Water activity (a_w) of meat
0	0	0·993
−1	2	0·990
−2	48	0·981
−3	64	0·971
−4	71	0·962
−5	74	0·953
−10	83	0·907
−20	88	0·823
−30	89	0·746

The freezing point of post-rigor muscle is in the range $-0.9°C$ to $-1.0°C$ (Winger and Pope, 1981). At most, only a very small fraction of muscle water is frozen at $-1°C$, but this fraction is substantial at $-2°C$, and increases rapidly with decreasing temperature to exceed 70% of total muscle water at $-4°C$. Further reductions in temperature produce progressively smaller increases in the frozen water fraction (Table II). Supercooling of meat may occur if temperatures fall to no more than a few degrees below $-1°C$, but most commercial freezing would aim for substantially lower temperatures (Jul, 1984). Supercooling is, therefore, unlikely to be important in frozen meat.

CHANGES INDUCED BY FREEZING OF MEAT

Ice formation in muscle greatly affects the water activity (a_w) of the tissue. The a_w of unfrozen muscle is determined by the colligative properties of the solutes present, but when tissue is frozen to any degree, the a_w becomes identical to that of pure water at the same temperature, irrespective of the quantities or qualities of solutes present (Fennema, 1981); the temperature dependency of a_w at subfreezing temperatures is shown in Table II. With prolonged frozen storage, ice may sublimate from some surface tissues causing the a_w in these regions to fall below the value determined by the ice vapour pressure. This freeze-drying process ultimately produces areas of desiccated tissue referred to as 'freezer burn' (Kaess and Weidemann, 1962).

Once freezing has commenced, continued cooling will produce progressive concentration of the muscle-tissue solutes, and of these, potassium and lactate are quantitatively the most important (Table I). Associated pH changes have been observed during the freezing of meat, and although there may be an initial small decrease in pH, at temperatures below $-5°C$, the pH rises approximately 0.3 units. Subsequent further small variations can occur during prolonged frozen storage (van den Berg, 1961; 1966).

COOLING RATES IN COMMERCIAL PROCESSES

As previously indicated, the rate of cooling defines the nature of the injurious processes to which micro-organisms will be exposed during freezing, and this determines, in part, the degree of damage inflicted.

With cooling rates of less than $1°C \, min^{-1}$, the intracellular freezing of microbial cells cannot occur, and it should be noted that more rapid cooling rates are unlikely to be obtained, even within small tissue masses, because of the insulating properties of the meat. Even meat surfaces, blast-frozen under normal commercial conditions, would appear to cool at rates significantly less than $1°C \, min^{-1}$; e.g. surface cooling rates of only $0.2°C \, min^{-1}$ have been observed with moderately severe freezing at $-30°C$ and an air speed of $1 \, m \, s^{-1}$ (James and Bailey, 1982). Only by the direct application of very cold liquids (cryogenic freezing) could the temperature at surfaces drop sufficiently rapidly for intracellular freezing to occur in comparatively large cells, such as in fungi, and the even faster rates ($>100°C \, min^{-1}$) required for intracellular freezing of bacteria are unlikely ever to be achieved in commercial practice.

EFFECT OF FREEZING ON THE MICROBIAL ENVIRONMENT

The quantitative and qualitative effects of freezing on a meat microflora will vary with the circumstances under which freezing occurs. The condition of the meat tissues, and the location of the micro-organisms in and on the tissues, can be the cause of differences in survival of equal, if not greater, significance than the variations between particular cooling processes. Such substrate-related variations may well occur within a single unit, such as a carcass. A heterogeneous response can, therefore, be expected from microfloras on meat undergoing commercial freezing, even when the same or similar cooling processes are applied.

During freezing, four specific meat environments can be identified.

1. *Dehydrated surfaces.* Here the microflora has been exposed to drying before freezing of the substrate. If solute diffusion to the underlying hydrated tissues is not possible, micro-organisms will simultaneously be exposed to increasing solute concentrations. This factor may, however, be relatively unimportant, as in such areas, e.g. carcass fat cover, the initial concentrations of potentially damaging solutes are likely to be much lower than corresponding concentrations in muscle tissue. Whether or not solute concentration accompanies drying at suprafreezing temperatures, freezing will merely reduce the temperature of the

dehydrated microbial environment, and should, therefore, have little or no effect on microbial survival at such sites.

2. *'Slow' freezing of moist exposed surfaces.* The freezing out of ice will occur first at surfaces directly exposed to low temperatures, and organisms at such sites are, therefore, likely to be trapped within this primary layer of pure ice and not exposed to elevated solute concentrations. With slow rates of cooling ($<1°C\,min^{-1}$), the only effect of freezing, apart from decreasing temperature, is progressive dehydration of the microflora.

3. *Fast freezing of moist exposed surfaces.* Micro-organisms tend to be trapped in frozen solution rather than pure ice, but undergo dehydration as in pure ice. Solute effects are likely to be small, as rapid freezing precludes substantial solute concentration. Cooling rates are unlikely to be sufficiently high to cause intracellular freezing of bacteria, but intracellular freezing of yeasts and moulds might occur if cooling rates after freezing approached $10°C\,min^{-1}$.

4. *Slow freezing within a tissue mass, e.g. comminuted meats.* Micro-organisms in this situation could be frozen within forming ice crystals or exposed to increasing solute concentrations. The majority of solutes present in meat are likely to be inert or even protective against freezing injury (Janssen and Busta, 1973), but lactate must be considered a likely cause of freezing injury. Lactic acid, like other organic acids, can have lethal effects on micro-organisms, the active entity being the undissociated acid. In the normal pH range of meat, i.e. 5·5 to 5·8, lactate is the greatly predominant species. The bacteriocidal activity of lactic acid is weak in comparison with that of acids such as acetic or formic, but, as with these other acids, the susceptibility of micro-organisms to lactate increases with decreasing pH and decreasing temperature. Consequently, most meat-associated species of bacteria have at least some strains that, at chilled temperature, are sensitive to lactic acid in the concentrations found in meat (Gill and Newton, 1982). In the early stages of freezing, micro-organisms will be exposed to increasing concentrations of lactate at a pH perhaps a little below that of the unfrozen meat which, in meat of normal ultimate pH, is low enough for bacteriocidal activity. This lethal effect would, however, reduce as the pH rises towards 6·0 at temperatures below −5°C.

FLORAS OF CARCASS MEATS

Site of Microbial Activity

It has been frequently claimed that deep muscle tissue harbours an 'intrinsic' flora derived from the gut via the circulatory or lymphatic systems (Ingram and Dainty, 1971; Rosset, 1982), but it is now clear that deep tissue from animals killed under reasonably hygienic conditions is usually sterile (Gill, 1979). Micro-organisms contaminate the surface tissues during dressing and processing. Portion cuts or other butchered meats acquire the same surface microfloras through spread from knives and other cutting implements. Surface bacteria do not penetrate into muscle tissue until high bacterial numbers have been attained and overt spoilage has developed (Gill and Penney, 1977). Micro-organisms on meat are, therefore, confined to surfaces unless comminution or tenderisation has destroyed tissue integrity and spread bacteria from the surface throughout the meat.

Initial Microflora

The initial contamination of carcass meats is derived mainly from the animal's hide or fleece (Empey and Scott, 1939; Nottingham *et al.*, 1974). The microflora of the hide or fleece originates from soil, water and vegetation as well as animal sources. The large majority of the organisms in the initial microflora are Gram-positive mesophiles, predominantly *Micrococcus, Staphylococcus* and *Bacillus* spp., originating from faecal material and the normal skin microflora. A small fraction are Gram-negative psychrotrophic species, mainly pseudomonads arising from soil, water and vegetation. Small numbers of enteric pathogens derived from faecal material may also be present. Butchering of carcasses adds to the microbial load as bacteria are transferred to meat surfaces from tools, work surfaces and hands. However, most of these contaminants are still Gram-positive mesophiles. Characteristically, red meats, dressed under hygienic conditions, bear an initial flora of 10^3 to 10^4 bacteria cm^{-2} of which less than 10%, and often only 1%, can grow at chiller temperatures (West *et al.*, 1972; Newton *et al.*, 1978). The extent of microbial contamination varies with the cleanliness of livestock presented for slaughter and the standards of dressing hygiene. The psychrotrophic content will also vary with season and geography as the number of psychrotrophic bacteria in soils falls markedly with increasing temperature (Grau, 1974). Yeasts and moulds are very minor components of the initial meat microflora. Species that can consistently be recovered are all common inhabitants of soil and vegetation, but are rarely present on meat in numbers exceeding $10^2\,cm^{-2}$, numbers less than this being characteristic.

The initial microflora of pork might be expected to differ from that of other red meats because dehairing of the skin by scalding or singeing reduces surface contamination to $<10^3$ organisms cm^{-2}, the majority of survivors being thermoduric. However, normal butchering procedures appear to recontaminate the pork meat with a microflora similar to that found on other red meats (Gardner, 1982).

Spoilage Microfloras

In developed countries, most meat is at least kept cool, and much of it is refrigerated. Refrigeration temperatures are too low to allow growth of mesophilic organisms, so, under such conditions, floras dominated by psychrotrophs develop. Even at higher temperatures, mesophiles grow too slowly to compete successfully with psychrotrophic species, and it is only when temperatures begin to approach the maximum for psychrotroph growth, about 30°C, that the latter organisms relinquish their dominance of meat spoilage floras (Gill and Newton, 1980*b*). There is little interaction between micro-organisms growing on meat until maximum flora numbers are approached.

Mesophiles, including pathogenic types, will proliferate only at temperatures above their minima for growth, and, although such organisms would normally make an insignificant numerical contribution to spoilage floras, public health considerations require that meat be held at temperatures of 7°C or less, such temperatures being taken as below the minimum for growth of most enteric pathogens (Angelotti *et al.*, 1959).

The spoilage floras of meats stored in air are almost invariably dominated by species of *Pseudomonas*. These strictly aerobic, Gram-negative organisms produce malodorous compounds when metabolising amino acids, and are the major cause of spoilage odour and off-flavour development in fresh meats. A second group of Gram-negative aerobes, the *Acinetobacter/Moraxella* group, may contribute substantial numbers to the floras of some meats of relatively high ultimate pH, but as they tend not to produce malodorous volatiles, they have only a low spoilage potential (Gill and Newton, 1978).

The facultatively anaerobic, Gram-negative organism *Alteromonas putrefaciens* is also pH sensitive, being unable to grow below pH 6·0, but, in contrast to the *Acinetobacter/Moraxella* group, it readily produces copious quantities of H_2S and organosulphur volatiles that rapidly render meat unacceptable. Growth of this organism can result in early spoilage of high-pH meat stored in air, or under anaerobic conditions as in a vacuum pack. Other facultative anaerobes, such as the Gram-positive *Brochothrix thermosphacta* and psychrotrophic species of the Gram-negative family Enterobacteriaceae, are associated mainly with spoilage of meats packaged anaerobically, as are the fermentative organisms of the genus *Lactobacillus* (Gram-positive). These latter organisms are of low spoilage potential, and their dominance of packaged-meat floras is essential for the extended shelf-life of chilled meat packaged under vacuum (Newton and Gill, 1978).

Because there is no species interaction, relative growth rates are the major factor determining domination of the spoilage flora, with the species that grow most rapidly under the particular storage conditions overgrowing competitors (Gill, 1985). However, the relative and absolute numbers at which different spoilage species are initially present can influence the composition of the final flora, and relatively high numbers of a slow-growing species can offset any growth-rate advantage of species present in relatively small numbers. Additionally, if absolute initial numbers of spoilage types are high, the number of generations elapsing before spoilage becomes evident is reduced, so that there is less opportunity for a growth-rate advantage to be expressed. Such effects can often account for flora differences between otherwise similar products stored under identical conditions.

RESPONSE OF THE MEAT MICROFLORA TO FREEZING

There are four distinct phases experienced by micro-organisms in foods that are frozen before consumption. These are (i) the chilling of cells and their environment to temperatures near 0°C; (ii) the freezing of the environment with an accompanying concentration of solutes; (iii) the storage of cells and their environment in the frozen state; and (iv) the thawing of cells and their environment. Any one of these phases may prove microstatic or microcidal for some or all species in a given microflora.

The effects of cooling and freezing on micro-organisms have been extensively studied and characterised for a limited range of model systems (Mazur, 1966; Ingram and Mackey, 1976), and from such studies, the overall qualitative microbial response to food freezing can be deduced. The quantitative effects of the various factors will, however, vary widely between individual species and, for organisms studied in simple systems, responses may be greatly modified by the composition of the complex menstrua that foods usually provide. Unfortunately, many of the published reports on the response of meat microfloras to freezing of meat fail to give any precise data for processing variables such as freezing rates, length of frozen storage and thawing rates, factors which are known to significantly affect microbial survival in model systems (Mazur, 1966). Discussion of microflora responses to meat freezing must, therefore, be limited to broad, qualitative responses, as currently available data do not permit detailed, quantitative evaluation of modifying influences attributable to the meat environment.

Chilling

All microbial growth is slowed as the environmental temperature decreases. However, the extent to which growth rates decrease with decreasing temperature varies between species and between strains of the same species. Despite this variation, for many bacteria the square root of the growth rate appears to be directly proportional to temperature over much of the growth range (Ratkowsky *et al.*, 1983). However, this simple relationship is not applicable at the extremes of the temperature range as the growth rate for any organism declines rapidly as temperatures fall within two or three degrees of its minimum. Growth of mesophilic organisms, with temperature optima around 37°C, is greatly reduced at 10°C, and most mesophiles will cease to grow when temperatures are only two or three degrees lower. By contrast, the temperature range for growth of psychrotrophs extends down to the temperature at which meat freezes.

For some organisms, the rapid transfer of growing cells to appreciably lower temperatures, within or below their growth temperature range, causes membrane barrier function to be lost, resulting in the release of intracellular components and an accompanying decline in the proportion of viable cells. This phenomenon is termed cold-shock (Marth, 1973). Susceptibility to cold-shock is more characteristic of Gram-negative than Gram-positive organisms (Ingram and Mackey, 1976). However, the cold-shock syndrome apparently occurs only in circumstances where sparse cell populations suspended in dilute media are cooled very rapidly from optimal growth temperatures to temperatures approaching the minimum for growth, and is precluded by rich media, dense cell suspensions and moderate shifts in temperature, or if temperature shifts are less than very abrupt (Gorrill and McNeil, 1960). Therefore, the conditions required for cold-shock could rarely, if ever, obtain for bacteria on the surfaces of meat undergoing freezing, and the possibility of cold-shock effects on meat microfloras can be discounted.

Freezing

Because of the diversity that exists amongst micro-organisms, even within narrow taxonomic groupings, it is difficult to predict with certainty the response of a given organism to freezing. Mazur (1966) has suggested that, at slow rates of cooling ($<1°C\,min^{-1}$), organisms that are tolerant of dehydration or immersion in non-penetrating solutes will be relatively resistant to freeze injury, whereas non-xerotolerant cells, whose growth is inhibited by moderate concentrations of external solutes, should have a marked susceptibility to freezing damage. The data available for the microfloras of meat tend to support this broad generalisation, with resistance to freezing damage tending to correlate with the ability to tolerate conditions of reduced water activity. However, the physiological changes that confer xerotolerance, notably the accumulation within a cell of solutes compatible with physiological activity and the maintenance of cell hydration in the presence of external water stress (Brown, 1976), will develop only under conditions of reduced water activity. Cells of an organism capable of some adjustment to conditions of water stress, but exposed to freezing in an environment of high water activity, may well have no opportunity to adjust to the stress imposed by freezing, and may, therefore, suffer greater freezing damage than they would if they had previously adjusted physiologically to an environment of low water activity. This aspect of the relationship between xerotolerance and resistance to freezing damage does not appear to have been examined, but it is possible that the responses of growing micro-organisms to freezing are significantly affected by the conditions they experienced in the period before freezing occurred.

Freezing exposes micro-organisms either to dehydration alone or to dehydration and elevated solute concentrations, depending on the location of the cells within the tissues. The extent to which microfloras are exposed to these two effects varies with the physical conditions applied to obtain freezing and with the state of the meat. For example, Schmidt-Lorenz and Gutschmidt (1968) observed that freezing killed a greater proportion of the flora of minced meat samples than of intact beef cuts. This difference could be ascribed to the bacteria distributed throughout the mince mass being more exposed to solute effects than organisms present only on the surface of meat cuts. However, such an effect cannot be reliably anticipated, as distribution throughout a mince would appear, in some circumstances, to protect against freezing damage (Gill and Harris, 1984). Moreover, the extent to which freezing is lethal for specific bacteria also depends on the physiological state of the organisms, growing cells being more susceptible than those in the stationary phase (Toyokawa and Hollander, 1956). Slow cooling would, therefore, tend to enhance the survival of mesophilic rather than psychrotrophic organisms, as the former would have ceased to grow before freezing commenced, while the temperature at freezing could still allow some growth of the psychrotrophs.

Bacteria vary widely in their resistance to freezing injury (Table III). In general, Gram-negative organisms are more susceptible to the lethal effects of freezing than the Gram-positive species, such as the micrococci and staphylococci, that commonly form substantial fractions of the initial microflora on meat (Haines, 1938; Empey and Scott, 1939; Kitchell, 1962; Georgala and Hurst, 1963). In contrast, vegetative cells of clostridial species associated with food poisoning are readily killed by freezing, although their spores, like those of most other spore-forming bacteria, are essentially unaffected by freezing (Barnes *et al.*, 1963). Of the Gram-positive organisms of importance in meat spoilage, data on *B. thermosphacta* appear to be lacking, while the lactobacilli that have been studied show a tolerance to freezing similar to that of Gram-negative organisms (Harrison, 1955). However, consideration of the diverse nature of the lactobacilli (Michener *et al.*, 1982) suggests that species responses to freezing could vary widely, whilst the moderate xerotolerance of *B. thermosphacta* would perhaps indicate some ability to resist freezing injury.

The response to freezing of Gram-negative organisms, both pathogenic mesophiles and psychrotrophic spoilage types, appears to be more uniform, and a reduction in numbers of between 90 and 95% can be expected for all types upon the freezing of meat (Georgala and

P. D. Lowry and C. O. Gill

TABLE III
Number of viable micro-organisms and percentage of total count in beef and raw minced beef before and after freezing at $-30°C$ (from Schmidt-Lorenz and Gutschmidt (1968).)

	Beef		Minced meat, raw	
	Before freezing	After freezing	Before freezing	After freezing
I Average no. micro-organisms g^{-1}	385 000	77 000	400 000	47 000
II % of total count				
1. Gram-positive bacteria	15	70	22	75
CBM-group (Coryne-, Brevi- and Microbacterium)	5	45	2	20
Gaffkya	—	21	6	15
Micrococcus	5	2	6	4
Leuconostoc	—	—	3	32
Lactobacillus	5	—	5	4
Bacillus	—	2	—	—
2. Gram-negative bacteria	85	30	78	23
Pseudomonas	75	22	69	23
Vibrio + Aeromonas	6	7	2	—
Achromobacter	4	1	7	—
3. Yeasts	—	—	—	2

Hurst, 1963; Foster and Mead, 1976). This reduction results in an 'enrichment' in the flora of the mesophilic Gram-positive species that predominate in the initial contamination (ICMSF, 1980a). These latter organisms play no part in the spoilage of chilled meat, so provided that chill temperatures are maintained during thawing, the lethal effect of freezing on Gram-negative spoilage organisms will somewhat extend the shelf-life of frozen/thawed meat over that obtainable for the same unfrozen meat stored under identical chill conditions. The reduction in numbers of Gram-negative pathogens, such as Salmonella, is also a beneficial effect of freezing, but, for hygiene purposes, it is necessary to assume that, if present, these organisms will always survive freezing in significant numbers.

In contrast to the above conclusions, it has been claimed that, during the freezing of packaged meat at $-4°C$ and even $-18°C$, numbers of psychrotrophic bacteria increase (Sulzbacher, 1950). However, much of the increase observed in that study can be attributed to microbial growth during slow cooling of the meat, as the insulation provided by the packaging would have retarded freezing; under these conditions, growth would have exceeded the subsequent lethal effects of freezing. It has also been claimed that yeasts and moulds, especially vegetative cells of the latter, are very sensitive to freezing and frozen storage (Borgstrom, 1955; Christophersen, 1968). However, at least some yeasts are resistant to damage by slow freezing rates and increased solute concentrations, and little damage is discernible at temperatures above −10°C; fungal spores behave similarly in aqueous suspension (Mazur, 1966). There are few quantitative data on the survival of filamentous fungi—partly because of the difficulty in defining 'cell survival' for these multicellular organisms—but the observation that outgrowth of germinated spores of psychrotrophic fungi is the same in frozen and unfrozen medium (Gill and Lowry, 1982), suggests that freezing and frozen storage at −5°C has little effect on the vegetative cells. In fact, yeasts and moulds on meat seem relatively resistant to freezing and frozen storage, as shown by there being no significant differences in fungal contamination between fresh and frozen meat (Oblinger and Kennedy, 1978; Karim and Yu, 1980).

Frozen Storage

Provided that temperatures for frozen storage remain below the minimum for growth of all elements of the microflora, no microbial proliferation can occur. The nature of the limiting conditions required to exclude microbial growth at freezing temperatures will be discussed in a later section, but no microbial growth is possible at 'normal' commercial frozen-storage temperatures, i.e. −12°C and below. In fact, extensive studies on pure cultures in frozen suspension indicate that microbial numbers will decline during extended frozen storage (Haines, 1938; Stille, 1950; Ray and Speck, 1973).

This loss of viability during frozen storage is far slower than that observed at the start of freezing, and the rate of loss decreases with time until cell numbers are essentially stable (Christophersen, 1968). Spoilage floras inoculated onto poultry meat in high numbers prior to blast-freezing and storage at −29°C (Kraft *et al.*, 1963) showed a pattern of population decline similar to that observed for Gram-negative organisms in pure cultures (Fig. 1). However, if substantial spoilage flora development occurs before freezing, the pseudomonads, which dominate meat-spoilage floras, will still form a significant proportion of the residual flora on frozen meats stored for prolonged periods (Zawadzki, 1972, 1973; Oblinger and Kennedy, 1978; Karim

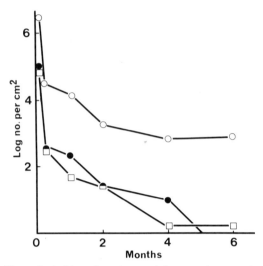

Fig. 1. The effect of air-blast freezing and frozen storage at −29°C on total aerobic numbers (○), fluorescing bacteria (●) and coliforms (□) on turkeys inoculated with a mixed flora. (Data from Kraft *et al.* (1963).)

and Yu, 1980). Those pathogenic organisms studied also decrease in number after prolonged frozen storage but, as with the initial effects of freezing, the complete elimination of pathogens cannot be expected (Georgala and Hurst, 1963). Spores are largely unaffected by prolonged frozen storage (Barnes *et al.*, 1963).

The holding temperature can affect the number of organisms surviving prolonged frozen storage. In model systems, the rate of inactivation decreases with decreasing frozen-storage temperatures, viability declining much more rapidly at temperatures of −5°C or above than at temperatures between −10 and −20°C (Haines, 1938; Ingram and Mackey, 1976). Specific data on the effect of storage temperature on spoilage organisms are lacking, but the survival of pathogens on frozen meat is also less at high than at low storage temperatures (Barnes *et al.*, 1963; Foster and Mead, 1976). Somewhat unexpectedly, fluctuation of temperatures within the freezing range (−20°C to −4°C) has been reported to cause numbers not to decrease below the levels observed for storage at the lowest temperature alone (Georgala and Hurst, 1963).

Thawing

Two types of injurious effect can be exerted on micro-organisms during thawing. In all cases, as the temperature rises towards the melting point of ice, micro-organisms can be exposed to solute effects in areas where early melting of the concentrated solutions occurs. In addition, where rapid freezing with intracellular formation of small ice crystals has occurred, both prolonged storage and thawing can allow ice recrystallisation with enhanced damage to the cell. Rapid transition through the thawing zone would limit the time of exposure to concentrated solutes and the opportunity for ice recrystallisation and, in fact, rapid thawing has been shown to be less destructive to frozen bacteria than prolonged thawing (Christophersen, 1968).

There appear to be no direct data on the effects of thawing alone on meat microfloras. However, from the previous discussion of freezing, it will be clear that intracellular freezing of meat micro-organisms is uncommon, and so subsequent related damage on thawing will be equally rare (Benedict *et al.*, 1961). Solute effects should only affect micro-organisms within tissue masses. Any such solute effects would be accentuated by the slow rates of thawing imposed by both the insulating effects of the meat mass, and the latent heat required for the melting of ice (Fennema and Powrie, 1964). In particular, the latent heat demand would tend to prolong the time at which organisms are held in the most harmful temperature range, a few degrees below 0°C.

GROWTH OF BACTERIA ON THAWED MEAT

Thawing of meat is a slower process than freezing. Thus, the poor thermal conduction properties of water, as compared with ice, mean that the thawing front takes about 20% longer to reach the thermal centre than the corresponding freezing front for equivalent temperature gradients. During thawing, the surface of the sample may rapidly approach the temperature of the thawing medium and remain at this temperature until thawing is complete, although with thick samples, surface temperatures are likely to remain below those of the surroundings for extended periods. When irregular-shaped items such as carcasses are being thawed, the different densities of meat over the carcass can lead to surfaces over the thinnest sections reaching, and subsequently remaining at, ambient temperature for considerably longer periods than surfaces over thicker areas (James *et al.*, 1977). Some microbial growth during thawing must, therefore, be considered likely, but the extent of any growth will be dependent on the mechanics of the thawing regime, the a_w and temperature of the meat surface, and the size of the meat sample.

Unfortunately, whilst the mechanics of different freezing regimes are well understood, and mathematical models for predicting thawing rates have been formulated (Bailey *et al.*, 1974), the microbiological consequences of the many thawing regimes are, in general, very poorly understood. There has been little attempt to determine the fundamental physiological responses of organisms to the conditions at the meat surface during freezing–thawing and immediately thereafter. Relevant data would include the significance of sub-lethal damage from freezing; the effect of freezing and frozen-storage time on lag phases; the relationship between thawing temperature and lag phase; growth rates and metabolic activity after freezing–thawing; the influence of drying and of pH; and the interactions of these variables. Without such data, responses of microfloras to specific thawing regimes cannot be accurately predicted.

Instead, the main approach to the thawing of meats has come from an engineering standpoint. Characteristically, these studies have examined specific thawing regimes with the aim of determining process capabilities, such as thawing time, meat appearance and weight loss; microbiological analysis being conducted only on a secondary basis (Vanichseni *et al.*, 1972; James *et al.*, 1977; Creed *et al.*, 1979). Compilation of the microbiological data from these studies leads to a piecemeal accretion of unrelated observations, and the lack of standardised environmental conditions within some studies further complicates interpretation. Because of this, it is almost impossible to draw any quantitative conclusions from existing data on the microbial consequences of thawing meat. At best, some limited comments can be made as to general trends.

Methods of Thawing

Commercial thawing of meats can be achieved by several methods: (i) air thawing by natural or forced convection, (ii) water thawing, (iii) vacuum thawing, (iv) dielectric thawing, (v) electric resistance thawing and (vi) microwave or infra-red thawing. The last three methods are

more suited to small, relatively homogeneous products where microbiological considerations are of no consequence, because of the rapid
warming rates. Vacuum thawing is not often used, as poor meat
appearance and high capital costs limit its application (Creed *et al.*,
1979). The heat-conduction methods, which use convected heat from
air, water or steam, are not, however, constrained by product shape or
size, and are, therefore, the most frequently used methods.

Water has better heat-transfer properties than air and, as expected,
immersion in water reduces thawing times compared with air.
However, there are associated problems (Vanichseni *et al.*, 1972;
Bailey *et al.*, 1974), for during immersion–thawing cycles, bacterial
loading of the thaw water can increase markedly, so necessitating
frequent water replacement. In addition, high standards of hygiene are
required to prevent the development of a heat-resistant microflora
associated with the recycle heating loop. It has been suggested that
meat thawed in water will have a 'wetter' surface than air-thawed
meat, and so will be more prone to subsequent spoilage (Bailey *et al.*,
1974). This is not always so, as the surfaces of air-thawed meats can
also be moist from the deposition of condensation when air flow is
slow. Air thawing is the most widely used method for thawing meats,
with the temperature, velocity and humidity of the air all influencing
the thawing pattern.

The effect of packaging on thawing rates is another area that has
received little consideration. The water-impermeable plastic packaging
films provide little insulation if closely applied to meat surfaces,
although insulation effects will occur where air gaps exist. Thawing
times will not, therefore, be significantly extended with shrink-wrapped
products, but the presence of the film will prevent surface drying,
which could otherwise inhibit bacterial growth.

Significance of the Lag Phase

During frozen storage, the organisms making up the meat microflora
will be in an induced stationary phase, provided that temperatures are
below their growth minima. Growth will, therefore, not begin
immediately upon reaching permissive temperatures during thawing.
Normally a lag phase will occur, its duration dependent upon the
organism itself, the temperature of thawing, and the microenvironment
at the meat surface. The duration of this lag phase could be very
important, as it is possible that, given the appropriate conditions,

thawing may be accomplished before the major components of the spoilage flora enter the exponential growth phase. Kitchell and Ingram (1956) found that when meat was thawed from $-20°C$, a psychrotrophic pseudomonad had a lag phase of approximately 10 h at 10°C, whilst Sulzbacher (1952) claimed that a similar organism had a lag phase of 2–5 days at 7°C. Whilst the latter observation indicates a longer lag phase than is commonly observed in most pure-culture studies (Arpai, 1963; Postgate and Hunter, 1963), a significant lag phase after thawing can be expected. Furthermore, there is evidence that the duration of this lag phase increases with increasing periods of frozen storage (Kitchell and Ingram, 1956).

In addition, model studies have shown that a proportion of micro-organisms experiencing freeze/thaw cycles will exhibit sub-lethal injury characterised by the differential ability to recover in minimal and nutritious media (Straka and Stokes, 1959; Ray and Speck, 1973). Organisms so injured have extended lag phases, presumably to undertake repair of cell damage (Souzu and Araki, 1962; Arpai, 1963). Meat is a sufficiently rich medium to permit recovery of injured cells, but extended lag phases due to freezing damage may still be anticipated after thawing.

Growth Rates of Bacteria on Thawed Meats

Historically, thawed meats were considered more susceptible to microbial spoilage than fresh meats (Borgstrom, 1955). This belief may have been, in part, a carry-over from observations on fruit spoilage, because freezing/thawing disrupts fruit tissue integrity and breaks the impermeable waxy skins, releasing juices that provide abundant substrates for microbial growth. With meat, however, although more drip is present on thawed than on fresh meat surfaces, the availability of low-molecular-weight substrates is not limiting to microbial development in either case. As the bacteria on the surface of chilled fresh meats grow at their maximal rates, no difference between growth rates on frozen/thawed or fresh meat can be expected. Experimental verification of this has been given by Kitchell and Ingram (1956) and Sulzbacher (1952).

As rates of flora development on thawed and fresh meat will be the same once growth has commenced, some predictions of the microbiological consequences of different thawing regimes can be made from examination of existing growth rate data for fresh meats. It must be

TABLE IV

Percentage composition of initial and final bacterial floras developing on unchilled beef. (Data from Gill and Newton (1980*b*, 1982).)

	Initial flora	Final aerobic flora		
		30°C	20°C	10°C
Pseudomonas	1	19	60	97
Acinetobacter	9	44	11	2
Moraxella	30	1	10	—
Micrococcus	52	—	8	—
Enterobacteriaceae	4	36	5	—
Lactobacillus	—	—	—	—
Clostridium	—	—	—	—
Brochothrix thermosphacta	4	—	5	1
Others	—	—	1	—

remembered, however, that any such predictions cannot take account of lag phases. *Pseudomonas* species dominate meat spoilage microfloras at temperatures up to 15°C by virtue of their more rapid growth rates (Gill and Newton, 1977), and at 20°C, the psychrotrophic pseudomonads still make up most of the microflora, although accompanied by substantial numbers of other psychrotrophs and some mesophiles (Table IV). As only the psychrotrophic pseudomonads increase markedly, in proportion to other genera, at low temperatures (<10°C), and these organisms form only a small fraction of the initial flora, thawing of meat in this cool temperature range might be expected to produce little change in the overall microbial loading, provided that extensive chill-flora development did not occur before freezing, and that thawing times are not excessive. Available data on air thawing of lamb carcasses and shoulders support this contention (Vanichseni *et al.*, 1972; Creed *et al.*, 1979).

Where warming curves are known for a specific thawing regime, the maximum number of generations for a given organism can be calculated to give an index of the microbial feasibility of the process (Table V). Klose *et al.* (1968) used this technique to determine whether unacceptable microbial growth (defined as 4–5 generations for a psychrotrophic pseudomonad) could occur during commercial thawing of turkeys over a range of temperatures. Whilst this approach is to be commended, it is disappointing that little effort was made to

TABLE V
Effect of temperature on the growth rate of bacteria on meat slices. (Data from Gill and Newton (1977, 1980*b*).)

Temperature (°C)	Generation time (h)				
	Non-fluorescent Pseudomonas sp.	Fluorescent Pseudomonas sp.	Enterobacter sp.	Acinetobacter sp.	Brochothrix thermosphacta
30	1·0	1·1	1·0	1·0	1·5
20	1·2	1·5	1·7	ND	1·4
15	2·0	2·0	2·4	3·1	2·8
10	2·8	3·0	3·5	5·2	3·4
5	5·8	5.4	7·8	8·9	7·3
2	7·6	8·2	11·1	15·6	12·0

ND, Not determined.

substantiate the predictive results with accompanying bacteriological analysis.

Temperatures for Thawing

Low-temperature thawing regimes (<10°C) are generally assumed to produce meats of better microbial status than equivalent processes at higher temperatures (Christophersen, 1968). This belief forms the basis of regulations, such as those implemented in France (Ministère de l'Agriculture, 1974), which state that the meat surface must not rise above 10°C for longer than 10 h during thawing. However, whilst the potential for microbial growth varies with different thawing regimes (Fig. 2), to date there is insufficient evidence to determine the relative merits of low-temperature/long-thaw versus high-temperature/short-thaw regimes.

The size of carcass or cut can greatly affect the microbial consequences of a given thawing regime. Significant increases in all components of the microflora have been observed on beef forequarters and hindquarters, but not on lamb carcasses, subjected to similar thawing regimes (James *et al.*, 1977; Creed *et al.*, 1979). The additional time that the surface temperature was at a level permissive for growth during thawing of the large-bulk beef samples was clearly the factor determining the extent of microbial proliferation. In these studies, only when large meat masses were being thawed did a significant correlation exist between increased microbial proliferation and increasing thawing temperatures, no significant differences being recorded for the

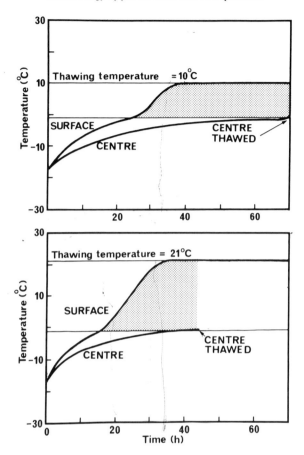

Fig. 2. Time–temperature profiles at the surface and centre of a 27 kg block of meat thawed at 10°C and 21°C, showing zones for potential microbial growth (stippled). (From Bezanson (1975).)

microflora of lamb carcasses thawed at temperatures ranging from 5°C to 20°C. These results further demonstrate the difficulty of applying the practical data obtained in a specific study to other, unrelated circumstances.

An additional consideration for thawing at temperatures above 10°C is the possible growth of pathogens. Inhibition of pathogen growth by the developing spoilage flora is unlikely (Gill and Newton, 1980*b*) but, as for the spoilage flora, data are lacking on the extent of the lag phase for pathogens on the meat surface after thawing. Assessment of the potential hazard from pathogens for specific high-temperature thawing regimes will require more complete data on the behaviour of pathogens in the post-freezing period.

Spoilage of Thawed Meat

From the above discussions, it is clear that thawed meat is as microbiologically sound as fresh meat, and as such will spoil in exactly the same manner for any given storage condition. The contention by Borgstrom (1955), that the longer frozen foods are stored the more rapidly they will spoil, has no basis. Rather, because of lower microbial contamination and the possibility of extended lag phases, the reverse is true. The rapid spoilage patterns reported for thawed meat may well have represented the development of unacceptable textural and chemical changes during freezing and thawing rather than specific microbial effects. Early adipose-tissue spoilage is, however, a possibility in thawed meats with exposed fat surfaces. Gill and Newton (1980a) demonstrated that, provided that fat surfaces are kept moist, spoilage can result at comparatively low bacterial densities ($10^6 \, cm^{-2}$) because glucose availability is limited at fat surfaces. As fat surfaces will become moist from deposition of condensation during thawing, and provided that subsequent storage does not allow drying, early spoilage of adipose tissue could well occur in thawed meat.

GROWTH OF BACTERIA ON FROZEN MEATS

Minimum Growth Temperatures

Spoilage by bacteria will not occur at storage temperatures below the minimum for their growth. The temperature range that encompasses the minimum for bacterial growth on meat has not been clearly defined, but psychrotrophic bacteria are certainly capable of growth below 0°C. Comprehensive reviews of the behaviour of micro-organisms at low temperatures cite numerous reports of bacterial growth at −5°C on meat, and as low as −12°C on microbiological media (Michener and Elliott, 1964; Schmidt-Lorenz, 1967). There is, however, reason to doubt that bacterial spoilage can occur even at −5°C, for much of the data on low-temperature growth pre-dates 1950, and was obtained under conditions giving no guarantee of strict temperature control. Because of the long periods of observation necessary to detect any increase in numbers when growth rates are very slow, the possibility of upward fluctuations of storage temperatures cannot be discounted (Michener and Elliott, 1964).

A further consideration is the effect of freezing. The minimum temperature for growth of cold-tolerant micro-organisms could be determined either by the physical effects of the medium freezing, or because the organisms, like thermophiles and mesophiles, have an intrinsic minimum temperature for growth. Since water activity of a frozen substrate is directly dependent on temperature, it is likely, as suggested by Scott (1957), that the simultaneous reduction in water activity with decreasing temperature sets physical limits for growth on frozen media. If water activity is the factor limiting growth of some organisms on frozen substrates, then growth at lower temperatures should be possible where supercooling occurs, as has been shown to be the case for some psychrophilic *Bacillus* species in laboratory media (Larkin and Stokes, 1968).

Leistner *et al.* (1981) have reviewed the extensive data pertaining to the water relations of micro-organisms associated with foods, and have compiled minimal a_w values for these organisms (Table VI). Some care must be exercised in interpreting the data, as species and strain variations can be expected for any genus. In addition, as most data were obtained under otherwise optimal conditions, the minimum water activity values for growth will increase as other growth-determining parameters approach limiting values.

Nevertheless, Table VI indicates that the bacteria primarily responsible for meat spoilage are not notably xerotolerant, having a_w minima between 0·98 and 0·95 at optimal temperatures. These a_w levels are attained in frozen meat between $-2°C$ and $-5°C$ (Table II). Freezing will, therefore, subject these bacteria to water stress at temperatures close to their minima for growth. The synergistic effect of the two factors probably ensures that the spoilage bacteria do not usually grow on frozen meat, i.e. below $-2°C$. The studies of Scott (1936) support this contention. He showed that growth of two *Pseudomonas* species was inhibited on ox muscle equilibrated to $0·98a_w$ at $-1°C$, and that growth of two *Achromobacter* species was marginal at $0·97a_w$ at the same temperature. (Most if not all of the strains referred to by Scott as *Achromobacter* would now be classified as non-pigmented pseudomonads (Brown and Weidemann, 1958).)

Consideration of water activity requirements might suggest that

P. D. Lowry and C. O. Gill

TABLE VI

Minimal a_w for multiplication of micro-organisms associated with foods. (From Leistner et al. (1982).)

a_w	Bacteria	Yeasts	Moulds
0·98	Clostridium[b], Pseudomonas[a]	—	—
0·97	Clostridium[c], Pseudomonas[a]	—	—
0·96	Flavobacterium, Klebsiella, Lactobacillus[a], Proteus[a], Pseudomonas[a], Shigella	—	—
0·95	Alcaligenes, Bacillus, Citrobacter, Clostridium[d], Enterobacter, Escherichia, Propionibacterium, Proteus, Pseudomonas, Salmonella, Serratia, Vibrio	—	—
0·94	Bacillus[a], Clostridium[e], Lactobacillus, Microbacterium, Pediococcus, Streptococcus[a], Vibrio	—	Stachybotrys
0·93	Bacillus[f], Micrococcus[a], Lactobacillus[a], Streptococcus	—	Botrytis, Mucor, Rhizopus
0·92	—	Pichia, Rhodotorula, Saccharomyces[a]	—
0·91	Corynebacterium, Streptococcus	—	—
0·90	Bacillus[g], Lactobacillus[a], Micrococcus, Pediococcus, Staphylococcus[h], Vibrio[a]	Hansenula, Saccharomyces	—

	Bacteria	Yeasts	Moulds
0·88	—	Candida, Debaryomyces, Hanseniaspora, Torulopsis	Cladosporium
0·87	—	Debaryomyces[a]	—
0·86	Micrococcus[a], Staphylococcus[i], Vibrio[j]	—	—
0·84	—	—	Alternaria, Aspergillus[a], Paecilomyces
0·83	Staphylococcus[a]	Debaryomyces[a]	Penicillium[a]
0·81		Saccharomyces[a]	Penicillium
0·79		—	Penicillium[a]
0·78	—		Aspergillus, Emericella
0·75	Halobacterium, Halococcus	—	Aspergillus[a], Wallemia
0·70	—	—	Aspergillus[a], Chrysosporium
0·62	—	Saccharomyces[a]	Eurotium[a]
0·61	—	—	Monascus (Xeromyces)

[a] Some isolates; [b] *Clostridium botulinum* type C; [c] *C. botulinum* type E and some isolates of *C. perfringens*; [d] *C. botulinum* type A and B and *C. perfringens*; [e] Some isolates of *C. botulinum* type B; [f] Some isolates of *Bacillus stearothermophilus*; [g] *B. subtilis* under certain conditions; [h] *Staphylococcus aureus* anaerobic; [i] *S. aureus* aerobic; [j] Some isolates of *Vibrio costicolus*.

some of the more xerotolerant, psychrotrophic, Gram-positive organisms could develop on meat at freezing temperatures. (Some strains of *Lactobacillus* have an a_w minimum of 0·90 and *Brochothrix thermosphacta* can grow at 0·94a_w (Grau, 1981).) However, direct evidence does not support this deduction. Recent studies confirmed an absence of bacterial growth on frozen lamb stored for a prolonged period at a constant temperature of $-5°C$ (Lowry and Gill, 1984a). In the only other relatively recent study on spoilage of frozen meat, chickens stored at $-2·5$, -5, $-7·5$ and $-10°C$ unequivocally supported bacterial growth only at $-2·5°C$ (Schmidt-Lorenz and Gutschmidt, 1969). The possibility that bacterial growth in this study may have occurred because the meat was supercooled rather than frozen can be discounted, as the meat was initially frozen to a temperature of $-30°C$. As supercooling of meat is improbable given usual storage conditions, the practical minimum growth temperature for meat-spoilage bacteria is probably in the range $-2°C$ to $-3°C$.

The growth of pathogens on meat at freezing temperatures can generally be discounted, as the most psychrotolerant strains of the common food-poisoning organisms, *Salmonella* and *Staphylococcus aureus,* have reported minimum growth temperatures of 6·7°C (Michener and Elliott, 1964). The one possible exception is *Yersinia enterocolitica*, which can develop slowly on raw meats at 0°C (Hanna *et al., 1977). However, there are no data indicating that this species can grow on frozen meat and, based on the pattern of other members of the Enterobacteriaceae, such growth can probably be dismissed on the basis of very low xerotolerance.

Enzyme Activity

It is often claimed that microbial enzymes, elaborated before frozen storage, remain active at freezing temperatures and contribute markedly to the organoleptic deterioration of frozen foods (Christophersen, 1968; Rosset, 1982). Whilst there is evidence to support the former claim (Rey *et al.*, 1969), the latter cannot be substantiated. Instead there are two lines of argument that tend to discount any such role of microbial enzymes. First, depending on the temperature of frozen storage, marked deterioration of the lipids in meat can result from the combined actions of tissue enzymes and autoxidation (Fennema, 1973a). At normal levels of microbial contamination on frozen meat, the microbial cell mass, and thus the quantity of microbial enzymes present, would be very small, so microbial enzyme activity would be insignificant compared with that of the muscle tissue. Secondly, bacteria grow at the expense of the low-molecular-weight substances on meat, and hence extracellular proteolytic and lipolytic enzymes are only elaborated as spoilage densities are approached (Glenn, 1976). Therefore, for microbial exoenzymes to play a role in organoleptic deterioration of frozen meat, near spoilage densities of bacteria would need to be present before freezing.

GROWTH OF FUNGI ON FROZEN MEATS

Although bacterial development is apparently restricted to growth on unfrozen substrates, the ability of certain yeasts and moulds to grow on frozen foods has long been recognised. In particular, mould spoilage of meat has received considerable attention; yeasts are not normally considered to play a significant role in meat spoilage (Walker, 1977). This predominance of literature relating to mould spoilage is undoubtedly due to the visible, and often spectacular, nature of the spoilage. Four main forms of mould spoilage are recognised in the meat industry. These are known by the descriptive terms 'black spot', 'white spot', 'whiskers' and 'blue–green mould'.

The most troublesome form is black spot, a spoilage condition associated particularly with frozen meat transported long distances by sea. The condition is characterised by uniformly black fungal colonies (3–8 mm diameter) whose hyphae penetrate the superficial layers of the tissue, and although meat so colonised presents no health hazard, it is aesthetically unacceptable. The meat can be trimmed, but this entails considerable financial loss due to the value of the meat removed, and the subsequent downgrading of carcasses and cuts (Thornton and Gracey, 1974). Initial reports concerning the causal organism of the condition indicated that a variety of moulds could be isolated from the black spots, and some doubts were even expressed as to its fungoid origin (Tabor, 1921; Wright, 1923). These observations were discounted by Brooks and Hansford (1923), who extensively surveyed the condition and concluded that the sole causative organism was *Cladosporium herbarum*. Their work has been accepted as definitive in all subsequent discussions of black spot spoilage. However, a recent study of the condition has shown that at least four different mould species, including *C. herbarum*, may produce black spot. A closely related species, *C. cladosporioides*, was found to be the predominant causative organism, with *Aureobasidium pullulans*, *Penicillium hirsutum* and *C. herbarum* being minor causal species. Colonies produced by the cladosporia spp. and *A. pullulans* were mainly subsurface, with the hyphae spreading along the intercellular junctions, possibly in response to the arid conditions at the meat surface; growth was superficial if the surface remained moist. Colonies of *P. hirsutum* remained on the meat surface, but they sufficiently resembled colonies of other species to be described as black spot (Gill *et al.*, 1981). The black spot moulds are all common species and cosmopolitan in distribution. They are abundant on aerial parts of plants, and it can, therefore, be assumed that they are transferred to meat from hides during carcass dressing procedures. The cladosporia have also been reported as common isolates from freezing works environments (Baxter and Illston, 1976), but the extent of contamination from these sources is unknown.

White spot spoilage often occurs in conjunction with black spot growths (Brooks and Hansford, 1923). The condition is characterised by the development of small, white, surface-restricted colonies (<4 mm diameter) that can vary in form from being woolly and tufted to sparse and flat. These colonies can be wiped off without leaving any visible evidence of growth. White spot spoilage was thought to be caused solely by *Sporotrichum carnis*, now *Chrysosporium pannorum* (Brooks and Hansford, 1923). A recent study has confirmed that *Ch. pannorum* is the predominant causal organism of this condition, but an *Acremonium* species also occasionally produces such spoilage. In addition, immature colonies of *Penicillium* species can produce similar spoilage forms (Lowry and Gill, 1984*b*).

The characteristics of whiskers and blue–green mould spoilage were investigated in this same study. In gross appearance, whiskers on meat is distinguished by its spreading, pale-grey, diffuse colonies with abundant aerial growth extending beyond the meat surface by up to 5–10 mm. This spoilage form is due principally to *Thamnidium elegans*, and occasionally to a second species, *Mucor racemosus*. The blue–green mould forms are distinguished by small, blue–green, woolly colonies (<3 mm diameter), *Penicillium corylophilum* being the only causal species isolated (Lowry and Gill, 1984*b*). Other *Penicillium* species with appropriate limits of xerotolerance and psychrotolerance may also be capable of producing such spoilage, e.g. Brooks and Hansford (1923) described *P. expansum* as another species isolated from blue–green moulds on meat. Like white spot mould growths, both whiskers and blue–green moulds form only superficial colonies on meat.

The gross appearance of the surface moulds can be markedly altered by the surrounding packaging. On naked or stockinet-wrapped carcasses, colony formation by the surface-restricted moulds proceeds unimpeded, but on carcasses or cuts shrink-wrapped in gas-permeable, moisture-impermeable plastic films, aerial growth is restricted. Under these conditions, growing hyphae spread diffusely across the meat surface, and only form visible colonies where the meat surface and the film are not in intimate contact (unpublished observations). In such cases, meat surfaces bearing no visible sign of mould growth may, in fact, support considerable ramification of certain moulds. Such considerations do not, however, apply to the black spot moulds, as most of their colony formation occurs in the sub-surface tissue, and therefore continues unhindered by wrapping films.

Minimum Growth Temperatures

Moulds are generally more tolerant of a decreased water activity than most meat-spoilage bacteria (see Table VI). Therefore, it is reasonable to expect that certain species of moulds could develop on frozen meat provided that they have a sufficiently low intrinsic minimum temperature for growth. Although the precision of much of the data is questionable, Michener and Elliott (1964) cite numerous reports of mould growth on meat at sub-zero temperatures, leaving little doubt as to the reality of mould growth on frozen meat. However, the practical limiting temperature for the growth of moulds on meat is less certain. Attempts to determine a minimum growth temperature, based solely on correlating the known a_w tolerance of an organism with the sub-zero temperature that would produce that a_w in frozen foods, have led to suggested growth minima as low as $-18°C$ (Leistner *et al.*, 1981). Such an approach overlooks the fact that psychrotrophs will have intrinsic minimum temperatures for growth; all micro-organisms examined in this respect having a limited temperature range, generally of some 35°C, outside of which growth cannot occur. In addition, unless the limits of xerotolerance of an organism are significantly greater than the corresponding a_w of a frozen substrate at the organism's minimum temperature for growth, its ability to tolerate arid conditions will tend to decline as the minimum growth temperature is approached. Therefore, the growth of some moulds on frozen substrates is likely to be inhibited, by a_w constraints, at temperatures above the intrinsic minimum for growth of the species.

Several reviewers of microbial behaviour at sub-zero temperatures have deduced that the minimum temperature for mould spoilage on frozen meat lies in the range $-10°C$ to $-12°C$ (Michener and Elliott, 1964; Thornton and Gracey, 1974; Ingram and Mackey, 1976). Such conclusions were reasonable on the basis of the data then available. However, much of the data concerning time and temperature requirements for the development of visible mould growth can be questioned on the grounds that, like the data for bacterial growth at sub-zero temperatures, the meat was stored under commercial conditions in which long-term accurate temperature control could not be assured.

Recent studies on the growth of black spot and other meat spoilage moulds at sub-zero temperatures indicate that the growth potential of these moulds has been greatly exaggerated (Gill and Lowry, 1982; Lowry and Gill, 1984b). Observations of the growth of individual fungal propagules in defined media under conditions of constant temperature showed that even the most psychrotolerant mould species had a temperature of between $-5°C$ and $-6°C$ as their practical minimum temperature for growth.

Meat spoilage moulds can be separated into three groups, based on their relative xerotolerance and minimum growth temperatures (Table VII). Three black spot species, C. herbarum, C. cladosporioides, P. hirstum, and the predominant white spot mould, Ch. pannorum, are all moderately xerotolerant and have a minimum growth temperature of approximately $-5°C$. This temperature undoubtedly represents the intrinsic minimum growth temperature of these species, as the a_w of a frozen substrate at $-5°C$ is well above their limits of xerotolerance. The recorded temperature minima for P. corylophilum, the Acremonium sp. and M. racemosus (-2 to $-1°C$) similarly represent intrinsic minimum growth temperatures despite the poor a_w tolerance of M. racemosus. These organisms are incapable of growth on frozen substrates. In contrast, the ability of T. elegans to grow at sub-zero temperatures appears to be limited by its poor xerotolerance. This organism grew at $-7°C$ in a supercooled medium, and extrapolation of the growth data indicated an intrinsic minimum temperature of $-10°C$ (Fig. 3). Although potentially capable of growth on supercooled meat at temperatures below $-5°C$, the unlikely occurrence of such supercooling suggests that, as for the other moulds, $-5°C$ defines the practical limits for growth of this organism.

TABLE VII

Minimum temperatures and water activities for growth of meat spoilage moulds. (Data from Gill and Lowry (1982) and Lowry and Gill (1984a).)

Organism	Mould spoilage form	Minimum growth temperature[a] (°C)	Minimum water activity[b] (a_w)	a_w of frozen meat at minimum growth temperature
Cladosporium herbarum	BS	−5·5	0·90	0·95
C. cladosporioides	BS	−5·0	<0·88	0·95
Penicillium hirsutum	BS	−5·0	<0·86	0·95
Chrysosporium pannorum	WS	−5·0	0·90	0·95
Acremonium sp.	WS	−2·0	<0·86	0·98
Penicillium corylophilum	BG	−2·0	<0·88	0·98
Mucor racemosus	W	−1·0	0·94	0·99
Thamnidium elegans	W	−5·0	0·94	0·95

BS, black spot; WS, white spot; BG, blue–green moulds; W, whiskers.
[a] Established on a nutrient medium containing glycerol to prevent freezing.
[b] Determined on a nutrient medium at 25°C with glycerol as the solute.

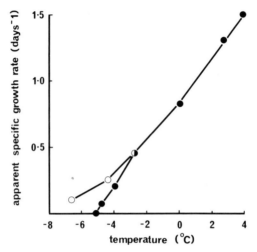

Fig. 3. The effect of water activity on the apparent specific growth rate of *Thamnidium elegans* at sub-zero temperatures: (●) glycerol-supplemented media at water activities equivalent to those of ice at sub-zero temperatures; (○) glycerol-supplemented medium of constant water activity, 0·98 (i.e., supercooled below approximately −2°C). (From Lowry and Gill (1984*b*).)

Observations similar to those for *T. elegans* have been made by Haines (1931*a*) for a strain of *Ch. pannorum* (which he called *Sporotrichum carnis*), which grew in a supercooled medium at −7°C and had an estimated minimum growth temperature of −10°C. Unlike the *Ch. pannorum* strain in the recent study, this organism was intolerant of reduced a_w, being unable to grow on frozen media at any temperature.

Growth Rates

Meat-spoilage moulds grow exceedingly slowly at sub-zero temperatures. Approximate minimum times for spores to develop into visible colonies at such temperatures can be estimated from their exponential growth rates (Table VIII). Assuming no lag phase and that spores had germinated to produce an initial hypha of 10 μm before freezing, the most rapidly growing black spot species, *C. herbarum*, would require at least one month to form a barely visible colony of 1 mm diameter at −2°C, and over four months at −5°C. These estimates are, in fact, excessively generous for a practical situation. In circumstances where

TABLE VIII
Apparent specific growth rates of meat-spoilage moulds at freezing temperatures. (From Lowry and Gill (1984c).)

Organism	Apparent specific growth rate $(days^{-1})$		
	$0°C$	$-2°C$	$-5°C$
Black spot			
Cladosporium herbarum	0·31	0·18	0·03
C. cladosporioides	0·22	0·12	<0·01
Penicillium hirsutum	0·37	0·16	<0·01
White spot			
Chrysosporium pannorum	0·35	0·17	<0·01
Acremonium sp.	0·10	<0·01	—
Whiskers			
Thamnidium elegans	0·67	0·50	0·03
Mucor racemosus	0·14	—	—
Blue–green mould			
Penicillium corylophilum	0·19	<0·01	—

germination occurred subsequent to freezing, or storage was at the lower end of the growth temperature range, there would be a lag phase of considerable duration. Such a period without growth would substantially increase the time for visible colony formation. Further, the rate of growth would slow with increasing colony size, as growth becomes linear when the diameter exceeds 0·2 mm (Plomley, 1959). These predicted rates for mould spoilage development have been confirmed in a prolonged storage trial at −5°C, where black spot colonies became visible only after 35 weeks' storage (Lowry and Gill, 1984a). At temperatures above 0°C, strains of *C. cladosporioides* demonstrate considerable variation in growth rates, but all strains have similar minimum growth temperatures (Gill and Lowry, 1982). Only *T. elegans* has a significantly faster growth rate than *C. herbarum* at sub-zero temperatures, developing nearly three times as rapidly at equivalent temperatures. However, even for this fungus, visible colony development on frozen substrates is very slow.

Conditions for Meat Spoilage by Moulds

As bacteria generally grow much more rapidly than fungi, mould spoilage of meat is thought to develop only when concomitant bacterial growth is inhibited. Temperature has usually been assumed to be the critical factor, with mould spoilage occurring typically on frozen meat. Practical circumstances where bacterial, but not fungal, growth is prevented by low temperature can be envisaged. If meat were stacked when only superficially frozen, the surface temperature of units in the centre would rise as they equilibrated to a uniform temperature, and the insulating effects of the outer layers could ensure that temperatures suitable for mould growth were maintained for long periods. The insulating effects of the surrounding meat could also create favourable conditions in meat stacked against uncooled bulkheads. However, as moulds grow extremely slowly at sub-zero temperatures, temperatures would have to be held in a very narrow range over extended periods for colonies to develop. This requirement makes it improbable that such conditions are those commonly responsible for the occurrence of mould spoilage.

Although the meat spoilage moulds grow much more rapidly at temperatures above 0°C, bacteria would normally dominate the microflora at such temperatures. However, given moderate surface desiccation, growth of the hygrophilic spoilage bacteria will be inhibited, and under these conditions the moderately xerotolerant moulds could produce significant visible spoilage in relatively short storage periods. It seems likely, therefore, that the moderately xerotolerant causal species of the black spot, white spot and blue–green mould spoilage forms usually develop as a result of meat surfaces reaching temperatures substantially above −5°C, with surface desiccation limiting growth of spoilage bacteria.

The predominant organism producing the spoilage form whiskers, *T. elegans,* appears to develop under rather different conditions. This organism is not significantly more tolerant of reduced a_w than the spoilage bacteria, and is, therefore, unlikely to develop under conditions where surface desiccation favours development of other mould forms. Instead, it is more likely that the comparatively rapid growth rate of this species at sub-zero temperatures allows it to compete successfully with bacteria to produce spoilage when meat is held near its freezing temperature (−2°C) (Lowry and Gill, 1984*b*). The growth-rate advantage of *T. elegans* over bacteria could be further enhanced by temperature fluctuations within the range 0°C to −5°C, as mould growth would continue throughout, but bacterial growth would be suppressed if the periods that the temperature was within their growth range were insufficient for resolution of their lag phases.

Therefore, in commercial practice, whiskers spoilage by *T. elegans* probably results from temperature fluctuations, which lead to periodic surface defrosting of the frozen meat with accompanying high humidity from condensation.

Yeasts

Yeasts normally play a minor role in the spoilage of red meats, both because they constitute only a small portion of the initial microflora, and because they grow slowly compared with most bacteria. Like the moulds, if yeasts are to become a major cause of spoilage, bacterial growth must be restricted. In general, yeasts are significantly more xerotolerant than spoilage bacteria, although as a group yeasts are not as xerotolerant as the moulds (Table VI). Yeasts would, therefore, be expected to develop on meats where surface desiccation is sufficient to inhibit bacterial growth, and dominance of microfloras by yeasts under such conditions has been observed in commercial practice. Law and Vere-Jones (1955) found that significant growth of yeasts of the genus *Mycotorula* (now called *Candida* (Lodder, 1971)) developed on shipments of chilled beef where temperatures were held in the range 0 to $-2°C$, and bacterial growth was retarded by control of the chiller humidity. Under controlled conditions, Scott (1936) showed that three asporogenous yeasts of the genera *Candida* and *Geotrichoides* (now called *Trichosporon* (Lodder, 1971)) had minimum a_w values on meat at $-1°C$ between 0·90 and 0·92, and that growth rates were only minimally affected by lowering the a_w to between 0·99 and 0·94.

The moderate degree of xerotolerance of these psychrotrophic yeasts, at temperatures only marginally above the freezing point of meat, suggests that several yeast species may be capable of growth on frozen red meats, but little consideration has been given to this possibility. Although Haines (1931*b*) observed that frozen lamb carcasses stored for prolonged periods at $-5°C$ developed a significant yeast microflora on exposed muscle surfaces, this finding seems to have been overlooked in all subsequent discussions of microbial growth on frozen meats. However, a recent study of microbial growth on meat at $-5°C$ confirms that yeast-dominated microfloras are likely to develop in these marginal freezing conditions (Lowry and Gill, 1984*a*). Maximal yeast numbers ($10^6\,cm^{-2}$) were reached after only 20 weeks at $-5°C$, the yeasts growing exponentially with a mean generation time of about 8 days (Fig. 4); in contrast, mould colonies did not appear until after 35 weeks. At maximal numbers, the yeasts formed discrete

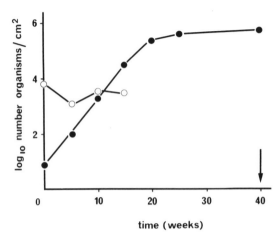

time (weeks)

Fig. 4. Changes in numbers of bacteria (○) and yeasts (●) during storage of frozen lamb at −5°C. The arrow indicates the storage time when mould colonies were first observed, the mean number of colonies per loin being 6. (From Lowry and Gill (1984c).)

pin-head colonies, but neither the yeasts themselves, nor their metabolic products, produced sensory changes in the meat (Winger and Lowry, 1983).

Four species of asporogenous yeast made up the microflora: *Cryptococcus laurentii* var. *laurentii, Cryptococcus infirmo-miniatus, Trichosporon pullulans* and *Candida zeylanoides,* the first mentioned being more than 90% of the population at all times. These same species have also been found in similar proportions in the microflora of chilled turkeys held for long periods at −2°C (Barnes *et al.*, 1978). The minimum growth temperatures for these yeast species have yet to be established, but in view of their likely limits of xerotolerance (0.90–$0.92a_w$), growth probably cannot occur beyond −8°C. Schmidt-Lorenz and Gutschmidt (1969) observed growth of a *Cryptococcus* species on frozen poultry at −5°C, and limited growth at −7·5°C was indicated; no growth was observed during prolonged storage at −10°C.

Yeasts as Spoilage Organisms

Although the extent of yeast growth on fat surfaces of lamb loins observed by Lowry and Gill (1984a) did not lead to gross colony formation or sensory deterioration of the meat, it is likely that, under some circumstances, a visible spoilage form will be produced on frozen meat. As spoilage numbers of bacteria are ultimately lower on fat than on muscle tissue, because of limited availability of substrates (Gill and Newton, 1980*a*), it seems likely that yeast numbers on the frozen fat surface would be similarly restricted. On muscle tissue, substrate limitation would not occur, and visible yeast colonies could be expected to develop eventually. At higher temperatures, any pre-formed yeast colonies would rapidly increase in size to give a more visible spoilage. Visible spoilage on fat may be enhanced by the deposition of haem pigments by the yeast, resulting in brown spots appearing on the surface (Egan and Shay, 1977). It is not clear, however, whether the species growing at freezing temperatures are identical with 'brown spot' organisms.

OFFALS

The term offals, or variety meats in American usage, is applied to a very diverse range of edible tissues. Some of these tissues, cheek and skirt meats, tongues and hearts, are muscular structures, and thus have characteristics that give patterns of microbial colonisation and spoilage similar to those generally observed for muscle tissues. However, offals also include organ and glandular tissues, such as liver, spleen and brain, where structure and composition are very different to that of muscle tissue. These differences can lead to significant variation in flora development between the two different types of offal tissue.

Unfortunately, there are few specific data on flora development on any offal except liver, the major offal traded internationally. However, the course of flora development on livers is probably paralleled in other organs of grossly similar structure and composition, i.e. tissues that are heavily vascularised and have a more open structure than muscle. The tissue is surrounded by a tough, smooth capsule of connective tissue, but because of the porous tissue structure, invasion of the deeper tissues by surface contaminants must be expected. Therefore, unlike muscular offals, organ offals can probably develop a spoilage flora in the deep-tissue environment.

Organ and glandular tissues are generally capable of high metabolic activities involving substantial protein turnover, and they can, therefore, be expected to undergo autolysis relatively rapidly compared with muscle. Despite their high metabolic potential, even organ tissues containing substantial glycogen reserves do not generally form lactic acid in the amounts that develop in muscle tissue as rigor progresses; the ultimate pH of organ tissue is usually well above 6·0. Consequently, pH-sensitive organisms that grow on DFD, but not on normal-pH meat, may commonly proliferate on organ tissues. In addition, as lactic acid is probably involved in the cell damage associated with solute concentration during the freezing of muscle, it is possible that such effects play a smaller part in microbial injury during the freezing of organ offals.

Microbiology of Fresh Liver

Chilled liver is generally believed to have a very short shelf-life. The usual commercial shelf-life cannot be estimated because specific data are lacking, but in one locality (Northern Ireland), chilled livers distributed for local retail sale are not expected to remain acceptable for more than three days after slaughter (Patterson and Gibbs, 1979). Why the shelf-life should be so short is not immediately obvious, but possibly: (i) livers are so heavily contaminated that the initial flora rapidly attains spoilage levels; (ii) liver composition is such that microbial spoilage becomes evident at low cell densities, and (iii) the treatment of livers is such as to allow bacteria to grow rapidly to spoilage levels.

The initial bacterial contamination of liver is reported to be $>10^5 \text{cm}^{-2}$ on the surface, and $<10^2 \text{g}^{-1}$ in the deep tissues (Gardner, 1971; Shelef, 1975; Rothenberg *et al.*, 1982). Much lower levels of contamination should be possible, as simple washing reduces bacterial numbers by at least an order of magnitude to give an average surface contamination of 10^2 bacteria cm^{-2} (Gill and De Lacy, 1982). The composition of the initial microfloras of livers can vary considerably but, in general, Gram-positive mesophiles of the genera *Bacillus, Micrococcus* and *Staphylococcus,* and coryneform bacteria dominate the flora. Representatives of the important spoilage organisms *Pseudomonas, Moraxella/Acinetobacter,* Enterobacteriaceae and *Brochothrix thermosphacta* are present, generally, as minor components, but in varying proportions (Gardner, 1971; Gill and De Lacy, 1982; Hanna *et al.*, 1982*a*).

The chill-spoilage microfloras of livers can be separated into three distinct, but variable, floras with respect to the availability of oxygen (Gill and De Lacy, 1982). The aerobic surface floras are characteristically dominated by *Pseudomonas, Acinetobacter* or a mixture of both; the 'anaerobic' deep tissues are dominated either by *Enterobacter, Aeromonas* or *Lactobacillus* and the drip, which will be partially anaerobic, is dominated by either *Pseudomonas* or *Enterobacter*. Typically, spoilage first becomes evident with the development of visible colonies on the surface, at which time the numbers of the deep tissue flora are some four orders of magnitude less than those of the surface flora. The deep-tissue flora, therefore, makes no significant contribution to the spoilage status of the product. Liver composition seemingly imposes little significant restraint on any species of the initial flora, so that the spoilage flora is both variable, and largely dependent on the relative numbers of the various psychrotrophic organisms in the initial flora.

Commercial Handling of Livers

Because of their supposed short shelf-life, livers, unless sold locally, are usually frozen. The livers are commonly packaged in plastic tubs while warm, sometimes individually, but more often in 2-kg bulk packs. After cartoning, the temperature in such tubs can be in excess of 30°C prior to chilling and freezing. Depending on the applied cooling conditions, it may take up to 10 h for the temperature at the centre of these tubs to be reduced to 10°C, and if the tubs are not chilled immediately, the period during which livers are incubated at warm temperatures will be correspondingly extended.

The closed tub provides an anaerobic environment for microbial growth. After 12-h incubation at 30°C, overall numbers can increase nearly 100-fold and, at this stage, although the mesophilic micrococci and staphylococci still dominate the flora, Enterobacteriaceae become a major component. These Enterobacteriaceae can comprise both psychrotrophic and mesophilic types, and the numbers of this group will continue to increase during chill storage; micrococci and staphylococci will die off. However, any chill flora will finally be dominated by lactobacilli, which grow faster than the Enterobacter-iaceae at chiller temperatures. Prolonged warm-temperature storage can result in floras composed almost exclusively of the potential pathogen, *E. coli*. In contrast, individual livers chilled without delay appear in excellent condition after storage for 5 weeks at 0°C (Gill and Penney, 1984) and a similar shelf-life has been observed for vacuum-packaged livers (Hanna *et al.*, 1982*b*).

Freezing of Livers

Clearly the temperature abuse characteristically inflicted on livers during commercial chilling and freezing will have a much greater influence on the microflora than the effects of freezing and thawing themselves. However, if extensive growth has not occurred during the freezing process, the initial levels of surface contamination and the approximate composition of the microflora are such that gross similarities between the behaviour of muscle and liver floras might be expected; a view supported by the limited data available (Hanna et al., 1982a). As with muscle tissues, there are no significant differences in rates or patterns of offal spoilage between thawed and fresh livers, either aerobically or anaerobically (Gardner, 1971).

Temperature abuse of offals during chilling and freezing favours the development of the mesophilic Enterobacteriaceae, and as this group includes pathogenic organisms, livers should not be held under conditions favourable to these organisms' growth. Specific pathogens, such as leptospires, may be present in some organs without overt disease symptoms (Ho and Blackmore, 1979); these could also proliferate during periods of temperature abuse. Any pathogens present on offals should respond to freezing similarly to their response on muscle tissue; their numbers would be reduced, but their elimination cannot be assured. Psychrotrophic organisms that will ultimately spoil the livers during chilled storage also proliferate when livers are kept warm, and any increase in their numbers during processing will ultimately be reflected in a shortened, post-freezing shelf-life.

Autolysis and Spoilage

A further consideration for the frozen storage of offals is the possibility of autolytic spoilage. Livers show a marked propensity for comparatively rapid autolytic spoilage at temperatures above 0°C, and chemical changes in fresh liver proceed 13 times faster at 37°C than at 15°C, and 3·5 times faster at 15°C than at 1°C; at 15°C, there is a lag of 1 to 2 days before autolysis becomes detectable. Appearance and texture changes associated with autolysis are undesirable, but do not render the tissue unacceptable. However, as autolytic products accumulate, livers will eventually develop an unacceptable bitter flavour due to the formation of peptides and/or tyrosine from degraded proteins (Rhodes and Lea, 1961). In fresh liver, these autolytic changes proceed comparatively slowly during chilled storage, and are only of minor importance by the time that gross microbial spoilage has developed (Gill and Penney, 1984). However, after frozen storage, autolytic activities are accelerated in the thawed liver, and provided that incubation temperatures exceed 10°C, autolysis proceeds three to four times faster in thawed than in fresh liver, because the lag period preceding the onset of detectable autolysis in fresh liver is eliminated (Rhodes, 1961). Repeated freezing and thawing of livers can, therefore, accelerate autolysis, and at the same time retard microbial growth, so that unacceptable autolytic changes may precede microbial spoilage.

Freezing of Other Offals

The limited data available on the compositions of other offal microfloras and their behaviour on freezing, suggest that little difference can be expected between these and liver microfloras (Hanna *et al.*, 1982*a*; Oblinger *et al.*, 1982; Rothenberg *et al.*, 1982). However, a precise knowledge of the behaviour of micro-organisms in these tissues still awaits analyses of the general microbiology of each specific meat.

MEAT PRODUCTS

Although there is wide diversity amongst meat products, most are shelf-stable at least at chilled and often at ambient temperatures. This is not surprising, as much current processing technology has been derived from traditional methods of preserving meat in the absence of refrigeration. Categories of product that can be considered stable include fermented meats, traditional cured meats of high salt content, intermediate-moisture foods (0·9 to 0·6a_w), and meats sterilised or pasteurised by cooking in sealed, retail containers. These products do not require freezing. In addition, freezing is often inappropriate for mild-cured products with salt contents insufficient to inhibit bacterial growth, as polyphosphate is usually added to enhance water binding; freezing causes a breakdown of the polyphosphate gel leading to extensive water loss and undesirable sensory changes. Even in the absence of polyphosphate, freezing of cured meats may be counter-productive as deteriorative chemical changes occur more rapidly at low than at high freezing temperatures (Bratzler *et al.*, 1977; Jul, 1984).

Therefore, the only meat products that are likely to be subjected to frozen storage are those containing fresh raw meat, and those that have been cooked but recontaminated during handling after cooking.

Raw, Comminuted Meat Products

Raw, comminuted meat products, e.g. fresh sausages, rissoles and hamburgers, reflect the microbial quality of the base meats. Where the meat ingredients have supported some microbial proliferation, a flora dominated by the normal Gram-negative spoilage psychrotrophs can be expected, the bacteria originally present on the surface of the meat being distributed throughout the comminute during processing. Sausage manufacture, in particular, is often associated with meats of poorer quality, e.g. rinds, trimmings from carcass meats, and meat remaining on carcasses after butchery (Sutherland and Varnam, 1982). Additives, such as breadcrumbs, rusks, and, in particular, spices, can also contribute to the flora of the finished product (Goto et al., 1971), but organisms introduced in this way will form only minor components of the flora. Fresh-meat products, without preservatives, characteristically develop Gram-negative spoilage floras similar to those of minced meat alone, the salt concentration being insufficient to limit their growth (Dyett and Shelley, 1966). If, however, sulphur dioxide is incorporated in the product emulsion, the onset of spoilage is delayed, and the microflora is dominated by Gram-positive organisms, particularly *Brochothrix thermosphacta*, lactobacilli and micrococci, with yeasts also being prominent (Dowdell and Board, 1971).

The responses of the various components of these microfloras to freezing will be similar to those for microfloras frozen in a fresh-meat mass. Although additional solutes, such as sodium chloride, are added to the meat emulsion, these are unlikely to add significantly to solute injury as they are present only in quantities sufficient to act as flavour enhancers. In fact, solute effects in these meat products are likely to be less severe than in a mince mass, as lactate, the principal solute in fresh meats, will be present in reduced concentrations because of dilution with the various admixtures. Therefore, although some reduction in bacterial numbers can be expected upon freezing, the shelf-life of previously frozen products will be similar to that of the fresh product, and will depend principally on standards of manufacturing hygiene.

Pasteurised Cured Products

Pasteurised cured meats encompass such diverse products as luncheon meats, meat loaves, patés and lightly cured sausages, such as frankfurters. These products are enclosed in sealed containers during cooking to internal temperatures between 60 to 70°C. If the containers remain intact, post-cooking bacterial contamination is prevented, and the spoilage flora can only arise from the surviving thermoduric *Bacillus,* group D *Streptococcus* and *Micrococcus* species. As these organisms generally cannot grow much below 10°C, such sealed products show marked stability during storage at chiller temperatures (Chyr *et al.*, 1980; Bell and Gill, 1982).

If the integrity of the casing or container is disrupted for the purposes of portion-slicing, post-cooking packaging or skinning, however, the surfaces of the product will adopt a microflora characteristic of the equipment used during handling (ICMSF, 1980*b*). Generally, a diverse microflora will be produced comprising both Gram-negative and Gram-positive species, and similar to that found initially on fresh meats (Blickstad and Molin, 1983). These products are often vacuum-packaged to extend their chilled shelf-lives for wide distribution, but for local usage, chilled storage alone will afford these products a useful shelf-life. In either case, the products will spoil, within a week or two, in a manner similar to that of chilled fresh meats. For long-term storage, this type of product must be frozen, and, on freezing, the microfloras, whether the initial, aerobic-spoilage types or the vacuum-package anaerobic-spoilage types, should behave similarly to the microfloras on fresh meat, and become frozen into an ice layer at the product surface. The response of the internal microflora, consisting of the surviving thermoduric species, is of little consequence, because the organisms are not involved in subsequent chill-temperature spoilage.

POULTRY

The Substrate

Poultry meat is the muscle tissue, attached skin and connective tissues of avian species, predominantly chicken and turkey, commonly used for food. Whilst there are many similarities between red meats and poultry meats, there are some important differences. In an entire poultry carcass, the intact surrounding skin serves as a physical barrier to micro-organisms that might otherwise contaminate the underlying muscle tissue. The skin, therefore, becomes the primary site of microbial activity on poultry carcasses, although mechanical defeathering leaves it with many holes and exposed feather follicles, which entrap contaminants during processing. Microtopographic studies have shown that the skin takes up a considerable amount of water during processing, and retains a fluid film containing serum proteins and other soluble compounds from which skin-bound micro-organisms seemingly derive their nutrients (Thomas and McMeekin, 1980). The pH of the skin is comparatively high and increases as chickens mature, being on average pH 6·6 for 9-week-old chickens and pH 7·2 for 25-week-old chickens (Adamcic and Clark, 1970).

Muscle tissue, too, is an important site of microbial activity, a large amount of chicken being sold as portion pieces, which have cut-muscle surfaces, or as chicken minces. Unlike red meats, the muscle tissue carries little intermuscular fat, most of the fat in poultry being found just under the skin and in the abdominal cavity. Muscle pH varies markedly between muscle groups, chicken breast muscle ranging from pH 5·7 to 5·9 while that of leg muscle is between pH 6·4 and 6·7 (ICMSF, 1980c). In other respects, muscle tissue composition parallels that of red meats.

Microflora

Before processing, poultry carry up to 10^5 micro-organisms cm^{-2} of skin, the majority of these being *Micrococcus* spp. and these must be regarded as the normal skin microflora as they are always present in high numbers and located at specific sites on the skin surface (Thomas and McMeekin, 1980). During processing, this predominantly Gram-positive microflora is largely removed, and replaced by a heterogeneous population composed mainly of Gram-negative bacteria. Although the feathers and feet of birds are contaminated with a diverse microflora derived from cage litter (Barnes, 1960), a large part of this flora is killed during scalding, and the skin of freshly plucked carcasses is relatively free of psychrotrophs (Clark, 1968; Thomas and McMeekin, 1980). However, recontamination occurs during feet and head removal, evisceration and chilling procedures, and process waters and ice, in particular, are major sources of psychrotrophic pseudomonads in the final floras (Clark and Lentz, 1969; Knoop *et al.*, 1971). Several authors have reported floras similar to those of red meats contaminating the skin of processed broiler carcasses (Barnes and Thornley, 1966; Daud *et al.*, 1979), but, in each study, the proportions of the component species vary. These differences appear to be related to changes in the proportion of the component types present in chiller water, and these fluctuations may occur on seasonal, day-to-day or locational bases (McMeekin and Thomas, 1978).

Pathogens

Poultry meat is a major source of human food poisoning worldwide, with chicken and turkey being principally responsible (Mead, 1982). The pathogens most frequently associated with poultry are *Salmonella* spp., *Clostridium perfringens* and *Staphylococcus aureus* (Hobbs, 1971). Studies of *Salmonella* contamination in processing plants have revealed that, although the organism is generally present in very low numbers, between 20 and 90% of carcasses may be contaminated during processing (Surkiewicz *et al.*, 1969; Van Schothorst *et al.*, 1975). Other enteric pathogens, *Campylobacter* and *Yersinia enterocolitica*, have been recovered with similar frequency from retail poultry (De Boer *et al.*, 1982; Kinde *et al.*, 1983), but the significance of poultry in their spread to humans has not been fully determined. This high incidence of contamination with enteric pathogens appears to have two major causes: (i) the practice of intensive rearing which encourages rapid transmission of pathogens through flocks; and (ii) the very high rates of throughput at large processing plants, which enhances the spread of micro-organisms among carcasses during processing.

Spoilage

The spoilage floras of chill-stored poultry develop in the same manner as red meat microfloras, psychrotrophic pseudomonads becoming dominant irrespective of their proportions in the initial flora. At moderate temperatures (15°C), the pseudomonads are displaced by other Gram-negative species, *Acinetobacter* and psychrotrophic Enterobacteriaceae, although they are still present in significant numbers at spoilage (Barnes and Thornley, 1966). The spoilage floras develop mainly on the skin surface, on cut-muscle surfaces under neck flaps, and in the feather follicles. Organisms are found in greatest numbers on the neck skin, the back and sites near the vent (Patterson, 1972).

Commercial Freezing Procedures

One of the major considerations determining the freezing regime applied to commercial poultry is the appearance of the frozen product. The colour of the frozen bird is greatly influenced by rate of freezing, and variations of the freezing rate produce any colour from dark red to chalky white. At slow rates of freezing, the large ice crystals allow incident light to penetrate the skin and muscle so that the dark-red deoxygenated myoglobin can be seen, but with rapid freezing rates, the ice crystals are smaller and reflect more light, so giving the surface a desirable white appearance. Rapid freezing is particularly important for birds scalded at 60°C, as this treatment removes the cuticle from the skin (de Fremery et al., 1977), and hence, it is the usual commercial practice to freeze chickens rapidly. Carcasses, which are protected from dehydration by plastic skin-films, are usually crust frozen to a depth of 2–3 mm either by air-blast freezing at temperatures as low as −34°C, or by an initial immersion-freezing or cryogenic spray freezing followed by air-blast freezing (Fennema, 1973b). The requirement for a chalky-white appearance is also a factor in determining frozen storage temperatures for poultry, for while the colour is stable at −29°C, it deteriorates through ice recrystallisation at −18°C within 5–6 weeks (Lentz and van den Berg, 1963). Frozen storage temperatures approximating −30°C are, therefore, commonly employed with poultry.

Response of the Microflora to Freezing

During commercial freezing, microfloras on poultry carcasses are exposed to two different environments. Those located on tissue surfaces within the body cavity would be likely to experience a slow freeze equivalent to that on red meats, as the insulating properties of the air space enclosed in the wrapped carcass would prevent rapid freezing. On the other hand, micro-organisms on the skin surface may freeze sufficiently rapidly to cause intracellular ice formation in large cells, such as yeasts and fungi. Whilst air-blast freezing and immersion-brine freezing are sufficiently rapid to cause intracellular crystallisation in tissue cells, such rates are unlikely to promote intracellular freezing in bacterial cells. The frequent occurrence of gross damage from intracellular freezing of yeasts and moulds would also appear doubtful, as Schmidt-Lorenz and Gutschmidt (1969) found that broilers, initially deep-frozen to $-30°C$, eventually developed a yeast-dominated flora when stored at $-5°C$. However, with cryogenic freezing, intracellular ice formation in yeasts and moulds seems certain. The possibility of intracellular ice damage in bacteria appears to be discounted by the findings of Kraft and Rey (1979), who observed no significant differences in counts of total aerobes or fluorescent pseudomonads for chickens frozen in liquid nitrogen or in air. Irrespective of the method of crust freezing, micro-organisms at the skin surface are likely to be frozen into a solution rather than a pure ice layer, because the time for solute migration during rapid ice-crystal growth is limited; such conditions should enhance freeze-injury in these micro-organisms because of solute effects.

Despite differences in freezing regimes, and likely differences in the microbial environments, available data indicate gross similarities between the behaviours of poultry meat microfloras and red meat microfloras during freezing. Viable counts on poultry decrease between 10 and 95% upon freezing (Kraft *et al.*, 1963; ICMSF, 1980*c*), and the Gram-positive species increase in proportion to the Gram-negative species during prolonged storage; the latter die off with time (Wilkerson *et al.*, 1961; Schmidt-Lorenz and Gutschmidt, 1969). Levels of *Salmonella* contamination decline during frozen storage, both at commercial holding temperatures (Davey *et al.*, 1976) and at higher temperatures ($-5°C$) (Foster and Mead, 1976), the rate of decline being more rapid at higher temperatures. However, it is clear that no commercial freezing process or cold-storage condition can be relied on to destroy salmonellae completely when these organisms are present in meat. Similar observations have been made with *Campylobacter*, up to 43% of a *C. jejuni* inoculum surviving frozen storage at $-20°C$ (Simmons and Gibbs, 1979; Norberg, 1981). As with red meats, thawed chicken meat exhibits the same spoilage characteristics as the fresh product even when thawing and refreezing is repeated (Elliott and Straka, 1964).

CONCLUDING REMARKS

Although prevention of microbial spoilage is the major reason for freezing of meat, the behaviour of micro-organisms in various commercial freezing and thawing regimes is not always a major factor in their design; instead engineering and economic considerations are usually paramount. This approach can give satisfactory results, but inadequacies must be expected in systems designed without reference to a parameter of prime importance. The inherent shortcomings of ignoring microbial behaviour are exemplified by the industry-wide acceptance of the abusive treatments employed for freezing of livers. Conversely, undue emphasis on particular aspects of microbial behaviour can arise from a failure to appreciate the multiplicity of factors that may affect microbial growth during commercial processing. An example of such misapplication of microbiological knowledge appears in hygiene regulations that permit thawing only at low temperatures, and ignore factors, such as product bulk and surface drying, that can be of greater importance for hygiene than process temperature.

There is good understanding of the general qualitative effects of freezing and thawing on micro-organisms, but detailed processing data are, unfortunately, scant. As a result, those considering meat-processing hygiene tend to ignore the combination of effects that can, in practice, ensure an acceptable microbiological status for products and, instead, identify from laboratory data values for specific factors that individually will preclude the envisaged hazards. Such oversimplified reasoning on hygiene matters is probably inevitable in the absence of data clearly relating processing variables to specific and quantified changes in the microbial flora of products. It is to be hoped that this data deficiency can be remedied relatively rapidly, so that substantial improvements in meat processing hygiene can be achieved with minimal costs and inconvenience.

REFERENCES

Adamcic, M. and Clark, D. S. (1970). *Journal of Food Science*, **35,** 103–6.
Angelotti, R., Wilson, E., Foter, M. J. and Lewis, K. H. (1959). *American Journal of Public Health*, **51,** 76–88.
Arpai, J. (1963). *Folia Microbiologica*, **8,** 18–26.
Bailey, C., James, S. J., Kitchell, A. G. and Hudson, W. R. (1974). *Journal of the Science of Food and Agriculture*, **25,** 81–97.
Barnes, E. M. (1960). *Proceedings of the 10th International Congress of Refrigeration, Copenhagen*, **3,** 97–100.

Barnes, E. M. and Thornley, M. J. (1966). *Journal of Food Technology*, **1**, 113–19.

Barnes, E. M., Despaul, J. E. and Ingram, M. (1963). *Journal of Applied Bacteriology*, **26**, 415–27.

Barnes, E. M., Impey, C. S., Geeson, J. D. and Buhagiar, W. M. (1978). *British Poultry Science*, **19**, 77–84.

Baxter, M. and Illston, G. M. (1976). *New Zealand Veterinary Journal*, **24**, 177–80.

Bell, R. G. and Gill, C. O. (1982). *Journal of Applied Bacteriology*, **53**, 97–102.

Bendall, J. R. (1973). In *Structure and Function of Muscle* (Ed. G. H. Browne), Vol. 2, 2nd edn, Academic Press, New York, pp. 243–309.

Benedict, R. G., Sharpe, E. S., Corman, J., Meyers, G. B., Baer, E. F., Hall, H. H. and Jackson, R. W. (1961). *Applied Microbiology*, **9**, 256–62.

Bezanson, A. (1975). In *Proceedings of the Meat Industry Research Conference*, American Meat Institute Foundation, Virginia, pp. 51–62.

Blickstad, E. and Molin, G. (1983). *Journal of Applied Bacteriology*, **54**, 45–56.

Borgstrom, G. (1955). *Advances in Food Research*, **6**, 163–230.

Bratzler, L. J., Gaddis, A. M. and Sulzbacher, W. L. (1977). In *Fundamentals of Food Freezing* (Eds N. W. Desrosier and D. K. Tressler), AVI Publishing Company, Westport, Connecticut, pp. 215–39.

Brooks, F. T. and Hansford, C. G. (1923). *Food Investigation Board Special Report No. 17*, HMSO, London.

Brown, A. D. (1976). *Bacteriological Reviews*, **40**, 803–46.

Brown, A. D. and Weidemann, J. F. (1958). *Journal of Applied Bacteriology*, **21**, 11–17.

Cassens, R. G., Marple, D. N. and Eikelenboom, G. (1975). *Advances in Food Research*, **21**, 71–155.

Christophersen, J. (1968). In *Low Temperature Biology of Foodstuffs* (Eds J. Hawthorn and E. J. Rolfe), Pergamon Press, Braunschweig, pp. 251–69.

Chyr, C. Y., Walker, H. W. and Sebranek, J. G. (1980). *Journal of Food Science*, **45**, 1732–5.

Clark, D. S. (1968). *Poultry Science*, **47**, 1575–8.

Clark, D. S. and Lentz, C. P. (1969). *Canadian Institute of Food Technology Journal*, **2**, 33–6.

Creed, P. G., Bailey, C., James, S. J. and Harding, C. D. (1979). *Journal of Food Technology*, **14**, 181–91.

Daud, H. B., McMeekin, T. A. and Thomas, C. J. (1979). *Applied and Environmental Microbiology*, **37**, 399–401.

Davey, G. R., Chiew, R. and Edwards, R. A. (1976). *Bulletin de l'Institut du Froid Supplement*, **1**, 211–18.

de Boer, E., Hartog, B. J. and Oosterom, J. (1982). *Journal of Food Protection*, **45**, 322–5.

de Fremery, D., Klose, A. A. and Sayre, R. N. (1977). In *Fundamentals of Food Freezing* (Eds N. W. Desrosier and D. K. Tressler), AVI Publishing Company, Westport, Connecticut, pp. 240–72.

164   *P. D. Lowry and C. O. Gill*

Dowdell, M. J. and Board, R. G. (1971). *Journal of Applied Bacteriology*, **34**, 317–37.
Dyett, E. J. and Shelley, D. (1966). *Journal of Applied Bacteriology*, **29**, 439–46.
Egan, A. F. and Shay, B. J. (1977). *Annual Report of the Commonwealth Scientific and Industrial Research Organisation (CSIRO) Meat Research Laboratory, Canon Hill, Australia*, p. 26.
Elliott, R. P. and Straka, R. P. (1964). *Poultry Science*, **43**, 81–6.
Empey, W. A. and Scott, W. J. (1939). Commonwealth Scientific and Industrial Research Organisation (CSIRO), Australia, Bulletin No. 126.
Fennema, O. R. (1973a). In *Low-Temperature Preservation of Foods and Living Matter*, (Eds O. R. Fennema, W. D. Powrie and E. H. Marth), Marcel Dekker Inc., New York, pp. 282–351.
Fennema, O. R. (1973b). In *Low-Temperature Preservation of Foods and Living Matter*, (Eds O. R. Fennema, W. D. Powrie and E. H. Marth), Marcel Dekker Inc., New York, pp. 504–50.
Fennema, O. (1981). In *Water Activity: Influences on Food Quality*, (Eds L. B. Rockland and G. F. Stewart), Academic Press, New York, pp. 713–32.
Fennema, O. and Powrie, W. D. (1964). *Advances in Food Research*, **13**, 219–347.
Fischer, C. and Hamm, R. (1978). *Proceedings of the 24th Conference of European Meat Research Workers, Kulmbach, West Germany*, **A6**, 1–6.
Foster, R. D. and Mead, G. C. (1976). *Journal of Applied Bacteriology*, **41**, 505–10.
Gardner, G. A. (1971). *Journal of Food Technology*, **6**, 225–31.
Gardner, G. A. (1982). In *Meat Microbiology* (Ed. M. H. Brown), Applied Science Publishers, London, pp. 129–78.
Georgala, D. I. and Hurst, A. (1963). *Journal of Applied Bacteriology*, **26**, 346–58.
Gill, C. O. (1979). *Journal of Applied Bacteriology*, **47**, 367–78.
Gill, C. O. (1985). *Advances in Meat Science*, Vol. 2, AVI Publishing Co., Westport (in press).
Gill, C. O. and De Lacy, K. M. (1982). *Applied and Environmental Microbiology*, **43**, 1262–6.
Gill, C. O. and Harris, L. M. (1984). *Journal of Food Protection*, **47**, 96–9.
Gill, C. O. and Lowry, P. D. (1982). *Journal of Applied Bacteriology*, **52**, 245–50.
Gill, C. O. and Newton, K. G. (1977). *Journal of Applied Bacteriology*, **43**, 189–95.
Gill, C. O. and Newton, K. G. (1978). *Meat Science*, **2**, 207–17.
Gill, C. O. and Newton, K. G. (1980a). *Applied and Environmental Microbiology*, **39**, 1076–7.
Gill, C. O. and Newton, K. G. (1980b). *Journal of Applied Bacteriology*, **49**, 315–23.
Gill, C. O. and Newton, K. G. (1982). *Applied and Environmental Microbiology*, **43**, 284–8.
Gill, C. O. and Penney, N. (1977). *Applied and Environmental Microbiology*, **33**, 1284–6.

Gill, C. O. and Penney, N. (1984). *Meat Science*, **11**, 73–7.
Gill, C. O., Lowry, P. D. and di Menna, M. E. (1981). *Journal of Applied Bacteriology*, **51**, 183–7.
Glenn, A. R. (1976). *Annual Review of Microbiology*, **30**, 41–62.
Gorrill, R. H. and McNeil, E. M. (1960). *Journal of General Microbiology*, **22**, 437–42.
Goto, A., Yamazaki, K. and Oka, M. (1971). *Food Irradiation*, **6**, 35–42.
Grau, F. H. (1974). *Advances in Meat Science and Technology*, **1**, Commonwealth Scientific and Industrial Research Organisation (CSIRO), Australia.
Grau, F. H. (1981). *Commonwealth Scientific and Industrial Research Organisation (CSIRO) Food Research Quarterly*, **41**, 12–18.
Hagen, P. (1971). In *Inhibition and Destruction of the Microbial Cell*, (Ed. W. B. Hugo), Academic Press, New York, pp. 39–76.
Haines, R. B. (1931a). *British Journal of Experimental Biology*, **8**, 379–88.
Haines, R. B. (1931b). *Journal of the Society of Chemical Industry*, **50**, 223T–7T.
Haines, R. B. (1938). *Proceedings of the Royal Society*, **B124**, 451–63.
Hanna, M. O., Stewart, J. C., Zink, D. L., Carpenter, Z. L. and Vanderzant, C. (1977). *Journal of Food Science*, **42**, 1180–4.
Hanna, M. O., Smith, G. C., Savell, J. W., McKeith, F. K. and Vanderzant, C. (1982a). *Journal of Food Protection*, **45**, 63–73.
Hanna, M. O., Smith, G. C., Savell, J. W., McKeith, F. K. and Vanderzant, C. (1982b). *Journal of Food Protection*, **45**, 74–81.
Hannan, R. S. (1975). In *Meat* (Eds D. J. Cole and R. A. Lawrie), Butterworths, London, pp. 205–22.
Harrison, A. P. Jr. (1955). *Journal of Bacteriology*, **70**, 711–15.
Ho, H. F. and Blackmore, D. K. (1979). *New Zealand Veterinary Journal*, **27**, 121–3.
Hobbs, B. C. (1971). In *Poultry Diseases and World Economy*, (Eds R. F. Gordon and B. M. Freeman), British Poultry Science, Edinburgh, pp. 65–80.
Ingram, M. and Dainty, R. H. (1971). *Journal of Applied Bacteriology*, **34**, 21–39.
Ingram, M. and Mackey, B. M. (1976). In *Inhibition and Inactivation of Vegetative Microbes*, (Eds F. A. Skinner and W. B. Hugo), Academic Press, London, pp. 111–151.
ICMSF (International Commission on Microbiological Specifications for Foods) (1980a). *Microbial Ecology of Foods: Vol. 1 Factors Affecting Life and Death of Microorganisms*, Academic Press, New York, pp. 1–37.
ICMSF (International Commission on Microbiological Specifications for Foods) (1980b). *Microbial Ecology of Foods: Vol. 2 Food Commodities*, Academic Press, New York, pp. 333–409.
ICMSF (International Commission on Microbiological Specifications for Foods) (1980c). *Microbial Ecology of Foods: Vol. 2 Food Commodities*, Academic Press, New York, pp. 410–58.
James, S. J. and Bailey, C. (1982). *Institute of Refrigeration Proceedings*, **78**, 33–41.

166   P. D. Lowry and C. O. Gill

James, S. J., Creed, P. G. and Roberts, T. A. (1977). *Journal of the Science of Food and Agriculture*, **28**, 1109–19.
Janssen, D. W. and Busta, F. F. (1973). *Applied Microbiology*, **26**, 725–32.
Jul, M. (1984). *The Quality of Frozen Foods*, Academic Press, London (in press).
Kaess, G. and Weidemann, J. F. (1962). *Food Technology*, **16**, 125–30.
Karim, M. I. A. and Yu, S. Y. (1980). *Malaysian Applied Biology*, **9**, 75–80.
Kinde, H., Genigeorgis, C. A. and Pappaioanou, M. (1983). *Applied and Environmental Microbiology*, **45**, 1116–18.
Kitchell, A. G. (1962). *Journal of Applied Bacteriology*, **25**, 416–31.
Kitchell, A. G. and Ingram, M. (1956). *Annales de L'Institut Pasteur de Lille*, **8**, 121–31.
Klose, A. A., Lineweaver, H. and Palmer, H. H. (1968). *Food Technology*, **22**, 1310–14.
Knoop, G. N., Parmelee, C. E. and Stadelman, W. J. (1971). *Poultry Science*, **50**, 530–6.
Kraft, A. A. and Rey, C. R. (1979). *Food Technology*, **33**, 66–71.
Kraft, A. A., Ayres, J. C., Weiss, K. F., Marion, W. W., Balloun, S. L. and Forsythe, R. H. (1963). *Poultry Science*, **42**, 128–37.
Larkin, J. M. and Stokes, J. L. (1968). *Canadian Journal of Microbiology*, **14**, 97–101.
Law, N. H. and Vere-Jones, N. N. (1955). Department of Scientific and Industrial Research (NZ), Bulletin No. 118.
Lawrie, R. A. (1974). *Meat Science*, 2nd edn, Pergamon Press, Braunschweig.
Leistner, L., Rödel, W. and Krispien, K. (1981). In *Water Activity: Influences on Food Quality* (Eds L. B. Rockland and G. F. Stewart), Academic Press, New York, pp. 855–916.
Lentz, C. P. and van den Berg, L. (1963). *ASHRAE Journal*, **5**(9), 42–6.
Locker, R. H., Davey, C. L., Nottingham, P. M., Haughey, D. P. and Law, N. H. (1975). *Advances in Food Research*, **21**, 157–222.
Lodder, J. (1971). *The Yeasts. A Taxonomic Study*. North Holland Publishing Company, Amsterdam.
Lowry, P. D. and Gill, C. O. (1984*a*). *Journal of Food Protection*, **47**, 309–11.
Lowry, P. D. and Gill, C. O. (1984*b*). *Journal of Applied Bacteriology*, **56**, 193–9.
Lowry, P. D. and Gill, C. O. (1984*c*). *International Journal of Refrigeration*, **7**, 133–6.
Marth, E. H. (1973). In *Low-Temperature Preservation of Foods and Living Matters*, (Eds O. R. Fennema, W. D. Powrie and E. H. Marth), Marcel Dekker Inc., New York, pp. 386–435.
Mazur, P. (1966). In *Cryobiology* (Ed. H. T. Meryman), Academic Press, London, pp. 213–315.
McMeekin, T. A. and Thomas, C. J. (1978). *Journal of Applied Bacteriology*, **45**, 383–8.
Mead, G. C. (1982) In *Meat Microbiology* (Ed. M. H. Brown), Applied Science Publishers, London, pp. 67–101.
Michener, B. J., Egan, A. F. and Roberts, P. J. (1982). *Journal of Applied Bacteriology*, **52**, 31–7.

Michener, H. D. and Elliott, R. P. (1964). *Advances in Food Research,* **13,** 349–96.

Ministère de l'Agriculture (1974). *Réglementation des conditions hygeniques de congélation de conservation et de décongélation des denrées animales et d'origine animales,* Article 20.

Newton, K. G. and Gill, C. O. (1978). *Journal of Applied Bacteriology,* **44,** 91–5.

Newton, K. G., Harrison, J. C. L. and Wauters, A. M. (1978). *Journal of Applied Bacteriology,* **45,** 75–82.

Norberg, P. (1981). *Applied and Environmental Microbiology,* **42,** 32–4.

Nottingham, P. M. (1982). In *Meat Microbiology* (Ed. M. H. Brown), Applied Science Publishers, London, pp. 13–65.

Nottingham, P. M., Penney, N. and Harrison, J. C. L. (1974). *New Zealand Journal of Agricultural Research,* **17,** 79–83.

Oblinger, J. L. and Kennedy, J. E. Jr. (1978). *Journal of Food Protection,* **41,** 251–3.

Oblinger, J. L., Kennedy, J. E. Jr., Rothenberg, C. A., Berry, B. W. and Stern, N. J. (1982). *Journal of Food Protection,* **45,** 650–4.

Patterson, J. T. (1972). *Journal of Applied Bacteriology,* **35,** 569–75.

Patterson, J. T. and Gibbs, P. A. (1979). *Meat Science,* **3,** 209–22.

Plomley, N. J. B. (1959). *Australian Journal of Biological Science,* **12,** 53–64.

Postgate, J. R. and Hunter, J. R. (1963). *Journal of Applied Bacteriology,* **26,** 405–14.

Ratkowsky, D. A., Lowry, R. K., McMeekin, T. A., Stokes, A. N. and Chandler, R. E. (1983). *Journal of Bacteriology,* **154;** 1222–6.

Ray, B. and Speck, M. L. (1973). *CRC Critical Reviews of Clinical Laboratory Science,* **4,** 161–213.

Reidel, L. (1957). *Kaltetechnik,* **9,** 38–40.

Rey, C. R., Kraft, A. A., Seals, R. G. and Bird, E. W. (1969). *Journal of Food Science,* **34,** 279–83.

Rhodes, D. N. (1961). *Journal of the Science of Food and Agriculture,* **12,** 224–7.

Rhodes, D. N. and Lea, C. H. (1961). *Journal of the Science of Food and Agriculture,* **12,** 211–24.

Rosset, R. (1982). In *Meat Microbiology* (Ed. M. H. Brown), Applied Science Publishers, London, pp. 265–318.

Rothenberg, C. A., Berry, B. W. and Oblinger, J. L. (1982). *Journal of Food Protection,* **45,** 527–32.

Schmidt-Lorenz, W. (1967). *Bulletin of the International Institute of Refrigeration,* **47,** 390–413; 1313–49.

Schmidt-Lorenz, W. and Gutschmidt, J. (1968). *Lebensmittel Wissenschaft und Technologie,* **1,** 26.

Schmidt-Lorenz, W. and Gutschmidt, J. (1969). *Fleischwirtschaft,* **49,** 1033–41.

Scott, W. J. (1936). *Journal of the Council for Scientific and Industrial Research,* **9,** 177–90.

Scott, W. J. (1957). *Advances in Food Research,* **7,** 84–129.

Scott, W. J. and Vickery, J. R. (1939). Council for Scientific and Industrial Research (CSIRO), Australia, Bulletin No. 129.

168 P. D. Lowry and C. O. Gill

Shelef, L. A. (1975). *Journal of Applied Bacteriology*, **39**, 273–80.
Simmons, N. A. and Gibbs, F. J. (1979). *Journal of Infection*, **1**, 159–62.
Souzu, H. and Araki, T. (1962). *Low Temperature Science Series B*, **19**, 49–57.
Stille, B. (1950). *Archive fur Mikrobiologie*, **14**, 554–87.
Straka, R. P. and Stokes, J. L. (1959). *Journal of Bacteriology*, **78**, 181–5.
Sulzbacher, W. L. (1950). *Food Technology*, **4**, 386–90.
Sulzbacher, W. L. (1952). *Food Technology*, **6**, 341–3.
Surkiewicz, B. F., Johnston, R. W., Moran, A. B. and Krumm, G. W. (1969). *Food Technology*, **23**, 1066–9.
Sutherland, J. P. and Varnam, A. (1982). In *Meat Microbiology* (Ed. M. H. Brown), Applied Science Publishers, London, pp. 103–28.
Tabor, C. J. (1921). *Cold Storage and Produce Review*, Oct. 20, 298–300.
Thomas, C. J. and McMeekin, T. A. (1980). *Applied and Environmental Microbiology*, **40**, 133–44.
Thornton, H. and Gracey, J. F. (1974). In *Textbook of Meat Hygiene*, 6th edn, Balliere Tindall, London, pp. 476–7.
Toyokawa, K. and Hollander, D. H. (1956). *Proceedings of the Society of Experimental Biology*, **92**, 499–500.
van den Berg, L. (1961). *Food Technology*, **15**, 434–7.
van den Berg, L. (1966). *Cryobiology*, **3**, 236–42.
Vanichseni, S., Haughey, D. P. and Nottingham, P. M. (1972). *Journal of Food Technology*, **7**, 259–270.
Van Schothorst, M., Northolt, M. and Kampelmacher, E. H. (1975). In *The Quality of Poultry Meat*, (ed. B. Erdstieck), Proceedings of the Second European Symposium on Poultry Meat Quality, Oosterbeek, The Netherlands, Spelderhoff Institute for Poultry Research, Beekbergen, Paper 24.
Walker, H. W. (1977). *Food Technology*, **31**, 57–65.
West, R. L., Berry, B. W., Smith, G. C., Carpenter, Z. L. and Hoke, K. E. (1972). *Journal of Animal Science*, **35**, 209–10.
Wilkerson, W. B., Ayres, J. C. and Kraft, A. A. (1961). *Food Technology*, **15**, 286–92.
Winger, R. J. and Lowry, P. D. (1983). *Journal of Food Science*, **48**, 1883–5.
Winger, R. J. and Pope, G. C. (1981). *Meat Science*, **5**, 355–69.
Wirth, F. (1976). *Fleischwirtschaft*, **56**, 988–94.
Wright, A. M. (1923). *Journal of the Society of Chemical Industry*, **42**, 488T–90T.
Zawadzki, A. (1972). *Medycyna Weterynaryjna*, **28**, 682–4.
Zawadzki, Z. (1973). *Medycyna Weterynaryjna*, **29**, 171–3.

Chapter 5

Microbiology of Frozen Fish and Related Products

C. K. Simmonds and E. C. Lamprecht

*Fishing Industry Research Institute, University of Cape Town,
South Africa*

INTRODUCTION

Although the practice of freezing fish has been employed in very cold
climates for centuries, and mechanical refrigeration has been applied
since the late nineteenth century (Cutting, 1955; Heen and Karsti,
1965), it was not until the second half of the twentieth century that
freezing began to acquire significance as a commercial means of
preserving fish. In recent years, however, it has become increasingly
important in replacing other more traditional means of preservation
such as salt-curing and drying. As shown in Table I, while the amount
of fish caught for human consumption has increased about threefold in
33 years since 1948, the quantity of fish frozen has grown to be almost
seventeen times greater in the same time. As the production of frozen
fish has assumed an even greater importance, so has the need for
knowledge of the microbiology of such products.

THE FREEZING OF FISH—SPECIAL
CONSIDERATIONS

Fishing is still largely a hunting occupation. Despite increasing
direction of efforts towards fish farming operations, these as yet only
account for some 5% of the world's total production of fish
(Cuddeford, 1983). The quality of the raw materials available is
consequently governed by the intrinsic properties of the various species

169

TABLE I

Proportion of world catch used for human consumption and processed as frozen fish (FAO, 1967, 1969, 1979, 1983)

Year	Total world catch	Proportion for human consumption		Proportion frozen	
	(Million tonnes)	(Million tonnes)	% of world catch	(Million tonnes)	% of world catch
1948	19·6	17·1	87·2	1·0	5·1
1958	33·2	27·9	84·0	2·8	8·4
1968	64·0	40·2	62·8	8·1	12·7
1978	70·4	48·2	68·5	15·4	21·9
1981	74·8	52·7	70·4	16·9	22·5

in their wild state: fish generally are neither bred to improve their processing characteristics nor treated to eliminate undesirable micro-organisms.

Fish and shellfish are, furthermore, cold-blooded organisms. They are conditioned by their environment, which, except for some surface waters in tropical areas, has a very much lower temperature than that obtaining in the flesh of mammals or birds. Both their indigenous micro-organisms and their enzymic systems are therefore still active at relatively low temperatures, with the result that the early application of adequate chilling and freezing assumes a critical importance in the maintenance of high quality.

THE MICROFLORA OF FRESHLY CAUGHT FISH

Early microbiological investigations were largely directed towards the safety aspects of fish as a food, using methods appropriate to the examination of meat for organisms capable of causing food poisoning. It soon became apparent, however, that such microbes are rarely found in fish caught in unpolluted waters and that spoilage of fish is caused by their naturally occurring microflora (Reay and Shewan, 1949; Liston, 1980). These are generally more or less psychrophilic and are capable of growing down to 0°C and, for some strains, several degrees lower (Hess, 1934a; Tarr, 1942). It is generally accepted that the flesh of freshly caught healthy fish is sterile (Tarr, 1942; Shewan, 1949). The skin and gills, however, may carry high loads of bacteria, as may the guts of fish which have been feeding.

Marine fish from cold or temperate waters have been found to have skin counts at 20°C of 10^1–10^7 cm^{-2}, gill counts of 10^3–10^7 g^{-1} and gut content counts of 10^3–10^8 cm^{-3}, (Shewan, 1962; Liston, 1980). Higher figures have been reported for Indian sardine (Shewan (1977), after Karthiayani and Iyer, 1967, 1971), but these counts were carried out at 30–37°C. While 37°C counts give lower recoveries than 20°C counts for cold-water fish (Liston, 1980; Simmonds and Lamprecht, 1980*a,d*), this may not be true for the flora from tropical species, which tend to be more mesophilic in their behaviour. For cold-water fish, counts incubated at 0°C are almost as high as for 20°C, and sometimes even slightly higher (Georgala, 1957).

The microflora on the skin and gill surfaces of cold-water marine species consist largely of Gram-negative rods such as *Pseudomonas*, *Moraxella* and *Acinetobacter* (formerly classed together as *Achromobacter*), *Flavobacterium* and *Vibrio*. The microflora on warm-water fish, on the other hand, numbers mainly Gram-positive types including *Micrococcus*, coryneforms including *Corynebacterium*, and *Brevibacterium* and *Bacillus* (Liston, 1980; Shewan, 1977). All these genera may, however, be found in varying proportions in the microflora on fish from both warm and cold waters (Shewan, 1962, 1977; Liston, 1980).

Evidence suggests that bacterial loads on skin slime and gills of fish show seasonal variations linked with changes in environment (Shewan, 1962). Skate and sole (Liston, 1955, 1956) and cod (Georgala, 1957, 1958) showed two peaks, in late spring and autumn, following on plankton blooms. On the other hand, a three-year study of the skin flora of hake caught within a 15-hour fishing radius of Cape Town, while showing considerable variations in both numbers and types of micro-organisms, did not reveal any seasonal pattern of variation (Simmonds and Lamprecht, 1980*c*, 1981, 1982). Such effects may have been obscured by other, long-term variations. For example, the dominant genus found on freshly caught hake in 15 out of the 17 catches examined in this study was *Pseudomonas,* whereas a few years previously *Achromobacter* was found to predominate (Simmonds and Lamprecht, 1980*b*). Similar variations have been noted by other workers. The major genus in the microflora of cod has been variously quoted as *Pseudomonas* (Georgala, 1957), *Achromobacter* (Thjøtta and Sømme, 1938), or *Micrococcus* (Dyer, 1947).

The microflora in the slime of fish has been stated to depend on the species of fish. Liston (1955, 1956) found markedly different bacterial loads on the skin, and especially gills, of skate when compared with sole caught in the same place at the same time. Wood (1953) showed teleosts caught in Australian waters to have *Micrococcus* as the dominant genus, while on elasmobranchs caught in the same area coryneforms predominated. Indian shark showed no *Pseudomonas* although the sea water in which they were caught contained large numbers of these organisms (Venkataraman and Sreenivasan, 1955). It is possible that specific antibiotic substances exist in the surface slime which suppress the growth of certain strains on a given species of fish. It should be noted that where striking variations have been observed between different types of fish, they often reflect differences between broad classes such as flatfish versus round fish, or cartilaginous fish as opposed to bony fish, rather than narrow inter-species differences.

There was no difference between the skin bacteria on two species of hake, *Merluccius capensis* and *M. paradoxus*, caught in the same waters (Simmonds and Lamprecht, 1980c, 1981, 1982). However, Georgala (1958) found significant differences between similar hake caught on the west (Atlantic Ocean) and east (Indian Ocean) coasts of South Africa. In the former, *Achromobacter* and *Pseudomonas* were in the majority, with lower numbers of *Corynebacterium* and other genera; in the latter, caught in warmer waters, *Corynebacterium* predominated, followed in order by *Achromobacter*, *Micrococcus*, *Pseudomonas* and others. Different species of fish from the North Sea exhibit smaller differences than those found between cod from the North Sea and Canadian waters (Scholes and Shewan, 1964). By and large, it is likely that the nature of the microflora on the skin and gills is governed more by the population of the water in which the fish is caught, than by the species of fish. Where such data are available for comparison, the correlation between the flora of sea water and that of fish caught therein can be quite striking (Shewan, 1977).

The numbers and types of bacteria recovered can also be influenced by the medium used and the incubation temperature. There has been some debate whether marine bacteria are really psychrophilic or even truly marine. Zobell (1946) expressed the opinion that they could not be described as psychrophiles because their optimum growth temperatures lay between 18°C and 22°C. The confusion really lies in the differing terminology used by different authors to describe psychrophiles and psychrotrophs. Several more or less conflicting definitions are mentioned by Jay (1978). His definition of psychrophiles, *viz.* organisms capable of growth at temperatures between 0°C and 7°C and producing viable colonies within seven days, is a convenient one.

Many marine micro-organisms will grow readily at −3°C and some strains will do so as low as −7·5°C (Hess, 1934a). While their most rapid growth is at about 20°C, the maximum crop size is obtained at around 5°C and crops at 0°C or −3°C are often larger than at 20°C (Hess, 1934b). They can survive almost indefinitely at −3°C to 5°C, yet many of them die quite rapidly at 37°C. Only 5% of the organisms recovered by incubation at 20°C from fish caught in the North Sea would grow at 37°C (Shewan, 1944) although 55% of those on fish caught off the Mauritanian coast would (Shewan (1977), after Kochanowski and Maciejowska (1964, 1969)). The spoilage bacteria of fish, especially from colder waters, can thus be regarded as psychrophilic.

Marine bacteria have been found to give higher counts on sea-water-based media than on those based on fresh tap water (Liston, 1956) and this has been reported even when enough sodium chloride has been added to the latter to make it isotonic with sea water, i.e. about 3% (Zobell, 1946). More recent studies have suggested that there is little difference in the counts obtained whether or not sea water is used in the media, provided that enough salt (usually 0·5%) is added to allow for the mildly halophilic nature of many marine organisms (Liston, 1980). The proportions of the different genera are, however, affected by the medium used. Liston (1957) found that the types of microflora recovered from skate and sole differed far more with the type of medium used (sea water or tap water) than with species of fish.

When Cape hake were stored in ice, no significant difference was found between total counts carried out in sea-water-based medium or in distilled water with 0·5% sodium chloride, at either 37°C or 20°C, although the 37°C counts on freshly caught fish tended to be higher on the former medium (Simmonds and Lamprecht, 1980a, 1980e). The counts obtained on a given medium at 20°C were, however, anything from 10 to over 80 times greater than those at 37°C. The proportions of the different genera recovered were affected by both the incubation temperature and the medium used. Figure 1 shows the percentage recovery of various types of bacteria from Cape hake in these trials. Regardless of medium, greater numbers of *Achromobacter*, *Pseudomonas*, *Brevibacterium* and *Corynebacterium* were found at 20°C, whereas *Micrococcus* was more prolific at 37°C and *Bacillus* was recovered only

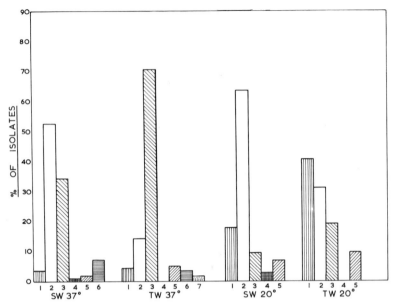

Fig. 1. Distribution of genera of bacteria on hake incubated on sea-water-based (SW) and distilled-water-based (TW) media at 37°C and 20°C. 1. *Achromobacter*, 2. *Pseudomonas*, 3. *Micrococcus*, 4. *Brevibacterium*, 5. *Corynebacterium*, 6. *Bacillus*, 7. *Flavobacterium*.

at 37°C. *Flavobacterium* was recovered in small numbers, only in distilled-water-based medium at 37°C. This probably reflects the general scarcity of *Flavobacterium* as a component of the microflora of Cape hake, rather than a preference for specific growth conditions. Regardless of temperature, greater crops of *Pseudomonas* and *Bacillus* were recovered from sea-water-based medium, while *Brevibacterium* was found only on sea-water medium. *Achromobacter, Micrococcus* and *Corynebacterium* grew more prolifically on distilled-water-based medium. Caution must be exercised in drawing conclusions from the results of experiments such as these. While the impression gained that *Micrococcus* and *Bacillus* are more strongly mesophilic than the other genera can be considered valid through extensive general experience with these organisms, the greater growth of *Micrococcus* in distilled water cannot be ascribed to a lower halophilism of this group. Marine micrococci are notoriously halotolerant and will grow on media such as skim-milk salt agar containing as much as 20% sodium chloride. The

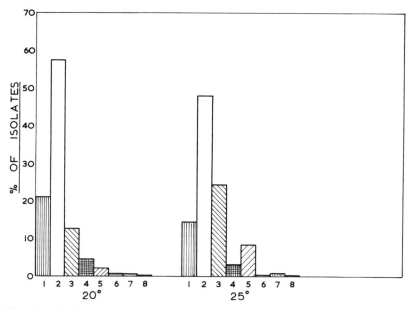

Fig. 2. Distribution of genera of bacteria on hake incubated on sea-water-based medium at 20°C and 25°C. 1. *Achromobacter,* 2. *Pseudomonas,* 3. *Micrococcus,* 4. *Brevibacterium,* 5. *Corynebacterium,* 6. *Bacillus,* 7. *Flavobacterium,* 8. Miscellaneous.

most probable explanation is that *Micrococcus* formed a greater proportion of the population owing to lack of competition from other organisms, particularly *Pseudomonas,* which have more stringent requirements for growth.

Even a temperature difference as small as that between 20°C and 25°C can cause significant changes in the proportion of genera recovered. Figure 2 shows the percentage recovery of different types of bacteria from Cape hake on sea-water medium at these two temperatures (Simmonds and Lamprecht, 1980c, 1981, 1982). Growth of *Achromobacter, Pseudomonas* and *Corynebacterium* was greater at 20°C, while *Micrococcus* and *Brevibacterium* were more prolific at 25°C. The preference of *Achromobacter* and *Pseudomonas* for lower temperatures, or their relatively greater ability to grow in the cold than other organisms is noted at 0°C and even lower. Georgala (1957) found that these groups form a higher percentage of isolates recovered from cod by incubation at 0°C than at 20°C, while for all other genera typed the reverse was true.

While the flora on the skin and gills may derive largely from the surrounding water, that in the digestive tract often resembles more closely that of the bottom mud. Anaerobes are usually absent from slime and gills, but are always present in the gut when fish have been feeding (Shewan, 1962), species of *Clostridium* often being found. Of special interest is *Clostridium botulinum*, one of the few human pathogens naturally occurring in the marine environment. The psychrophilic types E, F and non-proteolytic B have been found consistently in fish from localities where bottom sediments contain the enzymes (Shewan, 1977; Liston, 1980). They have been implicated, particularly type E, in outbreaks of botulism and can grow at temperatures as low as $3 \cdot 3°C$. It may be expected that this type of organism could cause problems in aquaculture systems, especially in natural earth ponds (Cann *et al.*, 1975). They must be considered a normal part of the flora in freshwater farmed fish, such as trout, even if in small numbers.

Freshwater fish tend to have lower counts than marine species, *viz.* $10^2–10^5 \, cm^{-2}$ for skin and $10^1–10^7 \, g^{-1}$ or ml^{-1} for gut content of temperate fish, and $10^3–10^5 \, cm^{-2}$ and $10^4–10^6 \, g^{-1}$ or ml^{-1} for skin and guts of tropical species (Shewan (1977), after various authors).

The microfloras of freshwater fish may include most salt-water genera, and in addition *Aeromonas*, *Lactobacillus*, *Alkaligenes*, *Streptococcus* and Enterobacteriaceae (Frazier, 1967). Warm-water fish often have large numbers of coryneforms and the Enterobacteriaceae may even include *Salmonella*. The kinds of bacteria are quite variable and are no doubt influenced by the population in the water. The changes in intestinal flora during the life cycle of Pacific salmon were studied by Yoshimizu and Kimura (1976). Freshwater, immature fish had 100% terrestrial bacteria in the gut, while transplant fish in 50% sea water had a 49/51% ratio of terrestrial to marine types. Ocean-dwelling fish had 100% marine bacteria, while adult freshwater spawners had 34% terrestrial and 66% marine types. Freshwater fish are more likely to be contaminated by faecal organisms and pathogens than marine fish, other than those caught inshore in areas close to sewage outfalls or other polluted streams. In addition to the organisms mentioned, there also appears to be a specialised freshwater microflora with anomalous characteristics and, as yet, the components are not readily classified (Bligh, 1971; Bramstedt and Auerbach, 1961).

Few data are available for shellfish straight out of the sea, most being for freshly landed fish. Figures for crustaceans from cooler waters range from 10^3–$10^7\,\mathrm{g}^{-1}$ at 20°C, with 37°C counts usually an order of magnitude lower. Tropical species caught in water around 20–25°C tend to have largely mesophilic floras, which give realistic counts when incubated at 30–37°C (Newell, 1973). Counts as high as 10^6–$10^7\,\mathrm{g}^{-1}$ have been obtained on shrimp directly out of the nets of trawlers in Malaysia (Cann, 1977). Counts for shellfish such as shrimp or prawn can furthermore be significantly affected by contamination by bottom sediment or contact with dirty deck surfaces (Iyer *et al.*, 1971; Liston, 1980).

Deep water molluscs such as scallops or queens usually have lower counts (10^3–$10^6\,\mathrm{g}^{-1}$ at 20°C) then those of estuarine species, such as oysters, cockles or mussels, which have given figures of 10^3–$10^8\,\mathrm{cm}^{-3}$ (Cann, 1977). The latter species are often examined for *E. coli* content only, as the waters in which they live are generally more subject to pollution.

The main groups of bacteria comprising the flora of crustacean shellfish are *Micrococcus*, coryneforms, *Achromobacter* and *Pseudomonas*, together with smaller numbers of *Flavobacterium/Cytophaga* and *Bacillus* (Cann, 1977). The proportional composition varies with the temperature of the water in which the animals live, cold-water species having largely *Pseudomonas* and *Achromobacter* (Liston, 1980), while *Micrococcus* and coryneforms dominate in crustaceans from warmer waters, e.g. Indian prawns (Sreenivasan, 1959). Gulf shrimp contained largely *Achromobacter*, *Micrococcus*, *Pseudomonas* and *Bacillus* (Williams *et al.*, 1952). *Micrococcus* predominated in the flora of Louisiana crab (Alford *et al.*, 1942), while *Achromobacter* was dominant in both flesh and gut of North Sea (Early, 1967) and Pacific crab (Lee and Pfeifer, 1975).

Molluscs caught in unpolluted waters probably have floras similar to those of crustaceans or fish living in the same area. Certain shellfish, however, when harvested from waters subject to contamination, can become a health hazard. Oysters and other bivalves feed by passing large quantities of water over a filter system, and the small particles, separated and retained in the alimentary canal, can carry with them organisms of faecal or other undesirable origin. Molluscs grown in sewage-contaminated water can be dangerous. The large outbreak of food poisoning due to oysters in New South Wales in 1978 provides a good example of the extent of this problem (Fleet, 1978). Contamination may be subject to periodic fluctuations, as for example after heavy rainfall, but in such cases, the bivalves may be self-cleansed by the process of depuration (Brown and McMeekin, 1977).

Apart from contamination by pollution, shellfish can carry food poisoning organisms of marine origin. *Clostridium botulinum* has already been mentioned; another widespread species is *Vibrio parahaemolyticus*. This is found on fish as well as shellfish, and has been responsible for upwards of 40% of the incidents of food poisoning in Japan (Kawabata, 1971; Barrow and Miller, 1976; Shewan, 1977) owing to the custom of eating raw or semi-preserved foods. The organism is sensitive to low temperatures, and for this reason is often not detected in sea water during winter months. It has been isolated from purified oysters, which had satisfied all the usual bacteriological tests, and it has therefore been suggested that marine bacteria should be used as an additional index of microbiological quality of shellfish (Barrow and Miller, 1969, 1972). Although *Vibrio* spp. are destroyed by heating, the accepted methods for cooking large shellfish, both molluscan and crustacean, are insufficient to do so (Jones, 1977).

A further class of micro-organism causing food poisoning in filter-feeding shellfish is the dinoflagellates, such as *Gonyaulex* and *Gymnodinium*. These are responsible for paralytic shellfish poisoning, produced by one of the most potent human toxins known (Ray, 1971). It is only partially destroyed by heat. The most effective means of prevention, as for other forms of contamination, is to ensure that sufficient time has elapsed after exposure to dinoflagellate blooms or 'red tides' for the molluscs to have purified themselves.

Moulds, yeasts and viruses do not play a large part in the microbiology of fish in so far as its quality as a food is concerned, but are connected mainly with fish diseases (Liston, 1980). *Rhodotorula* yeasts are occasionally responsible for pink discoloration in oysters.

THE SPOILAGE OF FISH

The spoilage of fish has been demonstrated to be caused mainly by bacterial activity. If fish flesh is sterilised by irradiation, however, the level of proteolytic activity due to the muscle enzymes is relatively low (Liston, 1965). Proteolysis proceeds very much more rapidly when caused by microbial growth. The enzyme activity produced when counts of $10^7 \, g^{-1}$ are reached is sufficient to cause autolysis in the absence of micro-organisms. Protozoan infection caused by myxosporidians, such as *Chloromyxum*, can also cause extremely rapid changes in texture in fish, such as hake or snoek (barracouta), when caught in warm waters and held at high ambient temperatures.

It has likewise often been shown that aseptically excised fish muscle kept under sterile conditions at chill temperatures does not putrefy during several weeks storage (Liston, 1980). The typical pattern of bacterial growth during the spoilage of fish follows the classical population curve, and is paralleled by well-defined stages in the deterioration of quality as defined by odour, flavour, appearance and other organoleptic factors. A typical spoilage sequence is that defined by Shewan (1977) for white fish such as cod or haddock. For the first six days in ice after catching, there is no marked spoilage, odours are virtually absent and the flesh is firm. This corresponds to the lag phase and the initial part of the logarithmic growth phase. For the next four days, the odour strengthens and becomes musty or mousey, and the flesh becomes softer. The logarithmic phase is now well under way. The third stage brings sour, sweet, bready, malty to fruity odours, and the flesh is definitely soft. Over these four days, growth is beginning to slow down. In the fourth and final stage, hydrogen sulphide and other sulphides are smelt, stale cabbage water, faecal and strong ammoniacal odours arise, and the flesh is soft and slimy. This stage parallels the resting phase; eventually numbers of bacteria begin to decline, but by this time the fish is long since completely putrid. At higher temperatures, the process is quicker, e.g. at 4°C the rate of deterioration is twice that of fish stored in ice.

The spoilage rate, and hence the shelf-life, also depends on the species of fish. For instance, the lag phase is generally thought to last for the duration of rigor mortis. During rigor, the pH of the muscle drops due to the accumulation of lactic acid. In cod the pH reaches 6·2 to 6·4 which is possibly marginal for the growth of spoilage bacteria, while in flatfish, especially halibut, it can drop as low as 5·5, which is considered inhibitory. Flatfish consequently have somewhat longer storage lives than round fish. It has also been suggested that this slower spoilage is due to the presence of larger amounts of lysozyme in the surface slime of flatfish, or that strains of bacteria causing spoilage are virtually absent from their skin flora (Shewan, 1977). The rigor condition of Cape hake is rather weaker than for cod, pH does not drop much below 7·0, and there are usually appreciable numbers of bacteria in the flesh at the time of landing freshly caught fish, i.e. after 12–18 h in ice.

Spoilage bacteria develop most rapidly in the gills, where the earliest signs of decomposition, such as off-odours, can be detected. They also grow in the surface slime on the skin and eventually penetrate into the flesh. The mechanism of penetration is not yet elucidated, and some controversy exists as to its nature. The slime is composed of mucoproteins and polysaccharides and contains free amino acids, amines and other compounds, forming an ideal substrate for bacterial growth.

Shewan (1961) outlines some of the biochemical changes occurring during spoilage of marine fish. These are dependent not only on the strains of bacteria present, but on the type of fish. For instance, the flesh of elasmobranchs, such as sharks and rays, contains large amounts of urea for osmoregulatory purposes, and this breaks down to ammonia early in the spoilage of these fish. Similarly, marine fish contain trimethylamine oxide (TMO), a substance lacking in fresh-water fish. Some strains of bacteria, mainly *Pseudomonas* can reduce TMO to trimethylamine (TMA), which is responsible for the so-called fishy odour in spoiling marine fish, and in larger quantities has a strong ammoniacal odour. The organoleptic quality of a fish is thus not necessarily related to the total count alone. Other *Pseudomonas* types considered particularly important as spoilage organisms are those producing sulphur compounds, such as hydrogen sulphide, dimethyl sulphide and methyl mercaptan, from the sulphur-containing amino acids cystine and methionine (Shewan, 1977). Fruity odours arise from the degradation of the amino acids glycine, serine and leucine to form esters of the lower fatty acids.

A further site for infection is the viscera, where bacteria may multiply, and their penetration into the flesh is assisted through digestion by gut enzymes. Removal of the guts, combined with thorough washing, can delay spoilage by this mechanism in fish such as hake, but other fish, including ocean perch, keep longer ungutted; this is due to spreading of the intestinal flora by evisceration.

Spoilage, especially at lower temperatures, is caused mainly by psychrophiles. The freshness of fish stored in ice, as assessed organoleptically, correlates well with total count at 20°C, but no such relationship exists for 37°C counts (Simmonds and Lamprecht, 1980a). Tables II and III show data for Cape hake stored at temperatures from 0°C (in ice) to 15°C in air (Simmonds and Lamprecht, 1980d). At all temperatures, 20°C counts increase rather more rapidly than 37°C counts. At 5°C and above, the 37°C counts correlate significantly with

TABLE II

Growth of bacteria on hake stored at different temperatures: 20°C and 37°C counts for trials 1–3

Storage temperature (°C)	Storage time (days)	Odour[a] rating	Total count (log_{10} organisms cm^{-2})[b]	
			20°C	37°C
Controls	0	8·3	4 913 (4 302–5 291)	3 861 (2 461–4 886)
15	1	6·2	6 665 (6 114–7 459)	4 819 (3 868–5 692)
	2	3·6	7 879 (7 579–8 413)	5 907 (5 324–6 779)
10	2	5·9	6 449 (5 218–7 281)	4 326 (2 946–5 539)
	3	4·2	7 413 (6 163–8 114)	4 687 (3 239–5 951)
5	3	6·4	6 531 (6 247–6 870)	4 598 (4 274–5 355)
	6–7	4·2	8 134 (7 563–8 465)	4 855 (3 743–5 956)
0	3–4	7·3	5 220 (4 502–5 577)	4 088 (2 401–5 634)
	6–7	6·3	6 488 (5 042–7 492)	3 607 (2 502–5 066)
	9–10	4·6	7 801 (7 218–8 373)	3 804 (2 380–4 976)

[a] Mean for 12 fish (3 trials × 4 fish each).
[b] Mean and range for 12 fish.

both storage time and freshness as assessed by odour on a 10-point scale (Rowan, 1956). Correlation coefficients for 20°C counts are significant at all temperatures, and generally at a higher level than for 37°C counts. *Pseudomonas* and *Achromobacter* predominate in 20°C counts, the former the more so with increasing storage temperature. *Micrococcus* is the dominant genus in 37°C counts, particularly for storage at 0°C. *Pseudomonas* is also recovered in fair numbers particularly at higher storage temperatures, and other genera including particularly *Bacillus* are found after storage at 15°C and incubation at 37°C.

The organisms responsible for spoilage are considered to be those which will produce off-odours when grown in pure cultures on sterile

TABLE III

Distribution of genera of bacteria on hake stored at different temperatures: 20°C and 37°C counts for trials 2 and 3

Storage temperature (°C)	Storage time (days)	20°C				37°C			
		Pseudomonas	Achromobacter and Alcaligenes	Micrococcus	Other	Pseudomonas	Achromobacter and Alcaligenes	Micrococcus	Other
Controls	0	53·7	3·8	35·0	7·5	5·0	0·0	91·2	3·8
15	1	65·0	25·0	10·0	0·0	10·1	2·5	68·4	19·0
	2	87·5	11·2	0·0	1·3	35·0	3·8	30·0	31·2
10	2	77·5	16·2	5·0	1·3	2·5	0·0	93·7	3·8
	3	62·5	35·0	2·5	0·0	16·3	0·0	81·2	2·5
5	3	25·0	61·2	13·8	0·0	6·4	2·6	85·9	5·1
	6–7	79·5	20·5	0·0	0·0	45·7	1·4	51·4	1·4
0	3–4	20·0	72·5	5·0	2·5	1·2	2·5	88·8	7·5
	6–7	37·8	43·9	14·6	3·7	6·3	0·0	93·7	0·0
	9–10	52·5	47·5	0·0	0·0	1·25	1·25	97·5	0·0

fish muscle or in its press juice. By these criteria, *Pseudomonas* and *Achromobacter* have been characterised as spoilers, and *Corynebacterium*, *Flavobacterium* and *Micrococcus* as non-spoilers (Lerke *et al.*, 1965*a,b*). It is generally accepted, largely on the basis of experiments with fish caught in cold waters in the northern hemisphere, that *Pseudomonas* dominate increasingly as spoilage progresses (Shewan, 1977; Liston, 1980). *Achromobacter* and *Flavobacterium* may have temporary spurts of growth, but are eventually overgrown; they persist, but in smaller relative numbers. Shewan's Group II *Pseudomonas* for example has been found to dominate after two weeks at 0°C (Jay, 1978, after Lee and Harrison, 1968, and Laycock and Regier, 1970). Cape hake, however, often still has a majority of *Achromobacter* as assessed by 20°C counts, after 9–11 days in ice (Fig. 3, histogram 2), and generally *Pseudomonas* and *Achromobacter* predominate together, in varying proportions, in the later stages of spoilage (Figs 3 and 4). *Achromobacter* and *Corynebacterium* have been reported to

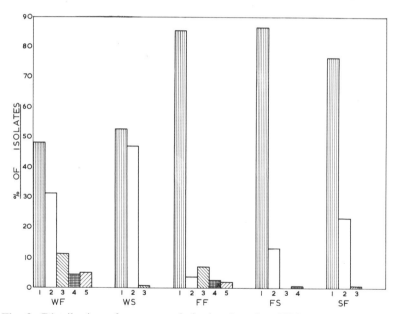

Fig. 3. Distribution of genera on hake incubated at 20°C on sea-water-based medium. WF = wet, fresh; WS = wet, stale; FF = shore frozen, fresh; FS = shore frozen, thawed and allowed to become stale; SF = stale when shore frozen. 1. *Achromobacter*, 2. *Pseudomonas*, 3. *Micrococcus*, 4. *Brevibacterium*, 5. *Corynebacterium*.

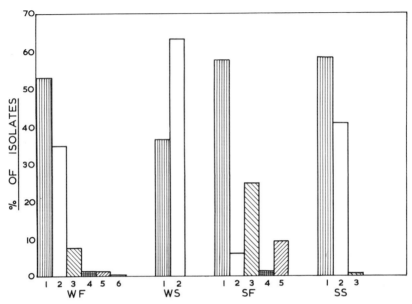

Fig. 4. Distribution of genera on hake incubated at 20°C on sea-water-based medium. WF = wet, fresh; WS = wet, stale; SF = sea frozen, fresh; SS = sea frozen, thawed and allowed to become stale. 1. *Achromobacter*, 2. *Pseudomonas*, 3. *Micrococcus*, 4. *Brevibacterium*, 5. *Corynebacterium*, 6. *Bacillus*.

predominate in shrimp spoilage (Fieger and Novak, 1961; Walker *et al.*, 1970; Cann, 1977). Spoiling crab meat has been found to contain *Proteus, Pseudomonas* and *Flavobacillus*, but Gram-positive organisms have also been implicated. The succession of genera and strains associated with spoilage at refrigerator temperatures still poses some unresolved questions.

The storage life in ice of marine fish from temperate waters varies with type, as previously mentioned. Round, lean fish, such as cod, haddock, hake and redfish, last from 8 to 15 days, while flatfish will keep for 17–21 days (Disney *et al.*, 1974; Shewan, 1977). Fatty fish have a somewhat shorter storage life, some 2–9 days, owing to the development of rancidity during spoilage. The storage life decreases with increasing temperature; for example, Cape hake which kept for 9 days in ice, lasted for 5, $2\frac{1}{2}$ and $1\frac{1}{2}$ days respectively at 5, 10 and 15°C. Total counts increase logarithmically during spoilage to maxima of 10^8–10^9 organisms cm^{-2} or g^{-1} on skin and flesh respectively.

Freshwater fish appear to follow a different pattern (Bligh, 1971; Disney *et al.*, 1974). Counts on surface slime may attain $10^{10}\,g^{-1}$ while flesh counts remain relatively low, and storage life in ice may range from 10 to 20 days. Tropical fish, owing to the relative scarcity of psychrophiles among their microflora, can be expected to have greater iced storage life than cold water species. Spoilage will not occur until the psychrophiles have multiplied to a sufficient level to produce off-odours. This is borne out in many instances, and the Indian perch, for example, has been found to have a storage life up to 45 days in ice, and West African marine fish 20 to 26 days (Disney *et al.*, 1974). Other factors are, however, also involved, as some species, such as the Indian pomfret, keep for as little as 7–9 days. Proteolytic enzymes, of either bacterial or muscular origin, play a greater role in the spoilage of tropical fish than of cold water species. Tropical fish can go off very rapidly under ambient temperature conditions; *Tilapia* are inedible within 15–20 h after death, although the total counts are only 10^3–$10^5\,g^{-1}$.

Similar considerations apply to the spoilage of shellfish. Tropical shrimp show a storage life of up to 16 days in ice, while cold water crustaceans and molluscs are spoiled after 8–10 days (Cann, 1977, after various authors). Total counts reach 10^7–$10^8\,g^{-1}$ at 20°C, with *Pseudomonas* or *Moraxella/Acinetobacter (Achromobacter)* predominating.

One major difference between the spoilage of molluscan shellfish, such as oysters, clams and scallops, and that of fish is due to the presence of large amounts of glycogen in the former. As a result of this, fermentative reactions form part of the spoilage mechanism, and this is reflected in a steady drop in pH. In fish, crustaceans and squids, on the other hand, the main result of initial spoilage is the production of large amounts of volatile basic nitrogen (Jay, 1978).

A problem associated largely with warm water fisheries is that of scombroid poisoning. It occurs mainly in fish belonging to the Scombridae, e.g. mackerel and tuna, and Carangidae, such as yellow tail, although salmon, pilchard and other fish have also been implicated. It is characterised by a wide variety of symptoms, among them a peppery taste, burning throat, thirst, itching, dizziness, headache, nausea, vomiting and diarrhoea. In almost every case, the implicated fish were found to contain high levels of histamine, often more than 100 mg per 100 g flesh (Liston, 1980). This is produced by bacterial action, especially where fish are exposed to high temperatures

for any length of time. The organisms concerned are able to
decarboxylate histidine, which is typically present in high levels in the
type of fish involved. Among the bacteria identified as responsible are
Proteus, *Klebsiella* and *Hafnia* (Omura *et al.*, 1978), while a wide
variety of other organisms possess the ability (Kimata, 1961; Ienistea,
1973).

Histamine itself, ingested at the level found in toxic fish, does not
produce the symptoms, but other amines, such as putrescine and
cadaverine, are usually present in large amounts, and it has been
suggested that they may also be implicated in the process (Kim and
Bjeldanes, 1979). While scombroid poisoning is not usually a threat to
life and recovery is rapid and complete, the intense discomfort can
severely affect sales of the products involved. The toxin is not heat
labile, and hot-smoked and canned fish have both been responsible for
outbreaks.

A form of skin irritation in workers handling, and particularly
gutting, affected fish has sometimes also been ascribed to histamine
poisoning. It is possible, however, that this condition is caused by
Erysipelothrix species, which have been repeatedly described as
infecting wounds or abrasions on the hands of fish handlers. It is not
clear whether these organisms are part of the natural flora of newly
caught fish, or introduced by contamination (Shewan, 1961).

THE FREEZING OF FISH

Fish may be frozen either at sea or on shore; the problems attendant
on the two types of operation differ to some extent. As soon as a fish
dies, various deteriorative changes start to take place, often very
rapidly at higher temperatures. It is usually desirable to bleed fish
before freezing, and the coagulation of fish blood is retarded by
lowering the temperature. Rigor mortis sets in soon after catching, and
the intensity and rapidity of the process can have a profound effect on
the quality of the fish (Jones, 1965). Rough handling of fish while in
rigor, and rapid passage through rigor at high ambient temperatures,
can cause severe damage to the texture, as evidenced by the gaping of
fillets cut from them. Fish caught in cold water can enter rigor in less
than an hour after catching if exposed to high ambient temperatures,
particularly if they have been caught by methods, such as trawling,
which leads to exhaustion before death, and can even pass completely

through rigor before a large catch is worked off into the fish room or the freezer. For these reasons, it is desirable to chill as soon as possible, ideally by discharging the net directly into a holding tank containing chilled sea water. A further argument can be made for rapid chilling when fish is to be held on board in the wet state, as the growth of spoilage bacteria is thereby retarded—two hours on deck can reduce the storage life in ice by a whole day.

It was originally thought that it was sufficient, when freezing fish, to lower the temperature just enough to halt bacterial degradation, say to −10°C. At this temperature, however, many undesirable chemical changes take place, resulting in a high quality storage life of only a few weeks, and virtual inedibility after a few months (Cutting, 1955). Toughening of the flesh due to protein denaturation proceeds rapidly, particularly in fish of the cod family; this is accompanied by excessive drip loss during thawing. Cold storage odours and flavours develop due to oxidation of the fats and complex interactions with the proteins, and fatty fish are especially prone to develop bitter rancid flavours, as the oils contain large amounts of highly unsaturated fatty acids (Heen and Karsti, 1965). Dehydration of the surfaces also take place, and even at −20°C these processes, although somewhat retarded, still proceed more rapidly than is desirable. It is generally recognised nowadays that frozen fish should be stored at −30°C or below if a high quality is to be maintained for a reasonable length of time. At this temperature, products will remain for all purposes as good as fresh for 4 to 8 months depending on species (Graham, 1982). Indeed, for some products such as the Japanese sashimi, which is consumed raw, it is common to maintain temperatures as low as −60°C to preserve the quality of fresh fish.

When fish are stored wet on-board the fishing vessel, prior to processing and freezing ashore, the changes due to bacterial action, even if retarded by chilling, reinforce the effects of poor cold storage and accelerate the development of off-odours and flavours. This is so even if the flavour of the fish, before freezing, was perfectly acceptable; with many species, a first class product cannot be obtained after more than about four days in ice. Again, exposure to elevated temperatures on the processing line hastens deterioration, and similar considerations apply to fish frozen whole or gutted, at sea or elsewhere, for thawing and further processing before re-freezing.

The method and rate of freezing also has an effect on product quality. Excessively slow freezing results in the build-up of more or less large ice crystals in the intercellular tissue, and the migration of water from within the cells leads to accelerated denaturation. By and large, modern freezing methods, such as plate and air-blast freezing, provide a sufficiently rapid freezing rate to ensure good textured quality. Freezing processes are often recommended which stipulate that the product should be removed from the freezer only when the temperature has been reduced to the point where, after equilibration, it will attain −18°C throughout. This should be taken as a minimum requirement; if at all possible freezing should proceed to an equilibration temperature of −30°C before cold storage. It is possible, however, that considerations of rapidity of freezing and ease of application can outweigh the advantages of the lower temperature. For example, it has long been the practice to freeze tuna intended for canning by direct immersion in brine. Usually the fish are first chilled to −1°C by refrigerated sea water, and the temperature then reduced further by dissolving salt in the water. This has the advantage, especially with large catches, that lowering of the temperature can be commenced rapidly and with comparatively little handling. Excessive salt penetration can be avoided by pre-chilling, low brine temperature during freezing, and separating the brine from the fish as soon as it is frozen.

In general, however, it must be accepted that fish and other aquatic creatures, such as molluscs and crustaceans, are more perishable and sensitive than most other foodstuffs, and must consequently be chilled immediately, frozen soon and rapidly, and held at as low a temperature as possible.

THE EFFECT OF FREEZING AND THAWING ON MICRO-ORGANISMS

Although some yeasts have been reported to grow at −18°C or even lower, and several bacteria and moulds have been found to grow at −12°C, it can be generally accepted that microbiological spoilage of fish is halted by freezing (Jay, 1978). It may proceed slowly at temperatures down to about −7°C, but in this range, loss of quality due to chemical reactions, such as oxidation of fats and denaturation of proteins, far outweigh the effects of bacterial action. At temperatures common in modern freezing practice, i.e. −18°C and lower, it is virtually non-existent.

The effect of freezing on the microbiological population is dependent on several factors. There is an initial mortality attendant on the freezing process itself. This varies with the type of organism, and seems to be virtually independent of the rate of freezing provided that the final temperature reached does not vary. This does not hold true in extreme cases, such as when fish is frozen very slowly by exposing it to a relatively high storage temperature of the order of, say, −7°C. The difference in mortality between different freezing methods found in normal practice, for example quick freezing by plate freezer and slower freezing in air at say −20°C, is however apparently negligible. Those cells which are still viable immediately after freezing die gradually during frozen storage, at rates dependent on the holding temperature. Mortality is most rapid at temperatures just below the freezing point, and less so with decreasing temperature—usually being slow below −20°C (Jay, 1978). Thawing and refreezing has a greater impact than single freezing, and for this reason, large fluctuations in holding temperatures cause a more rapid decline in numbers than more uniform conditions of storage.

The initial reduction in numbers of bacteria immediately after freezing can range from only one or two per cent up to 90 per cent (Shewan, 1961; Simmonds and Lamprecht, 1980b). The subsequent die-off is rapid at first, becoming slower as time passes; this can have the effect of lessening the effect of storage at higher temperatures. Thus, fresh Cape hake stored at −7°C suffered a considerably greater reduction in counts after one month storage than did those stored at lower temperatures (−18°C and −29°C). After two months, however, counts were substantially similar, on average, at all these temperatures (Simmonds and Lamprecht, 1980b). The effect of freezing and frozen storage is less pronounced on fish which have undergone spoilage; there seems to be some degree of protection afforded by the greater numbers, possibly because of the clumping of cells. The effect is, however, as with all too many phenomena in microbiology, rather variable.

Gram-negative organisms are generally more sensitive to freezing than Gram-positive ones, while spores and food poisoning toxins are unaffected by freezing (Shewan, 1961; Jay, 1978). *Pseudomonas* species are especially susceptible; this can have a profound effect on the spoilage rate of thawed, frozen fish. *Salmonella* and Enterobacteriaceae are among the more sensitive types, but the response varies greatly from strain to strain (Raj and Liston, 1961). The cocci, including marine micrococci and the enterococci, are more resistant to freezing. The coliform count, and especially that of *E. coli* on frozen prawns, has been found to drop by 95% or more during freezing and frozen storage (Mathen *et al.*, 1965; Lekshmy *et al.*, 1969), while the faecal streptococci were relatively unaffected. For this reason, it was suggested that the latter group be preferred as indicators of factory hygiene. Coliforms can, however, sometimes survive long periods at freezing temperatures (Lamprecht and Elliott, 1971). *Vibrio parahaemolyticus* has been found to be sensitive to low temperatures and dies-off rapidly (Lamprecht, 1980). *Staphylococcus aureus* has also been found to die-off during frozen storage, rather rapidly at first but more slowly as time progressed, so that significant numbers survive after several months (Lamprecht and Ferry, 1981). With high initial infections, the survival was such that compulsory hygiene standards could still not be met. Thus, although there is both a general and a selective mortality in bacterial populations during freezing and cold storage, it cannot be expected to improve on unsatisfactory microbiological practice to the point of acceptability. In fact, the best freezing practice—rapid freezing followed by storage at non-fluctuating low temperature—has the least effect on the micro-organisms. It is safest, therefore, to assume that freezing merely preserves the microbiological status quo. The effect of freezing on the naturally occurring microflora of Cape hake is illustrated in Figs 3 and 4. The proportion of *Pseudomonas* in shore frozen, fresh fish (less than a day in ice) is drastically reduced by freezing, while that of other groups with the exception of *Achromobacter* is diminished to some extent. The last-named genus is the most resistant to freezing and overshadows all others in the frozen fish. Sea frozen fish, compared with fresh, wet fish caught in the same area at the same time of year, show a similar decrease in relative numbers of *Pseudomonas*. The proportions not only of *Achromobacter* but also of *Micrococcus* and *Brevibacterium* are, however, somewhat higher than for the freshly landed wet fish. This is accounted for by the fact that, even after only 18 h in ice, the *Achromobacter* and *Pseudomonas* have been outgrowing the other genera to an extent sufficient to alter the composition of the microflora significantly from that at the time of catching.

The spoilage of thawed frozen fish has been shown to be slower than that of fresh wet fish (Shewan, 1961). Nevertheless it has recently been stated that no evidence exists that thawed products spoil more or less quickly than fresh ones (Liston, 1980). This is a result of several factors which may tend to tip the balance in favour of either fresh or thawed fish. Firstly, the numbers of bacteria are reduced by freezing as is the proportion of pseudomonads, the major group of odour-producing spoilage organisms. On thawing, the surviving organisms are suffering to some extent from both thermal and dehydration shock, resulting in an extended lag phase. On the other hand, damage to the fish tissues due to freezing, thawing and cold storage denaturation, results in the release of substrates, and enzymes which may degrade larger molecules, resulting in an overall increase in utilisable nutrients. This in turn leads to a potentially greater growth rate during the logarithmic phase. The method of thawing may influence the results. Thawing in water may remove surface loads of bacteria, but it may increase them if recirculation is practised. Thawing in air at high velocity and low humidity may dehydrate cells on the surface of the fish, and temporarily make nutrients less available by drying out. Finally, it is difficult, in practice, to thaw fish without the exterior surfaces being exposed, sometimes for hours, to elevated temperatures that are ideal for bacterial incubation. This tends to obscure the direct effects of freezing and thawing on spoilage.

Figures 3 and 4 show how species of *Pseudomonas*, having been considerably reduced in numbers by freezing, fail to attain as high a proportion of the total microflora as in wet fish even after storage in ice to the point of becoming unacceptably stale. In fact, as shown in Fig. 3, the *Pseudomonas* population of stale fish is not much affected by freezing, and is proportionately higher than for fresh-frozen fish allowed to become stale after thawing.

Figures 5–12 show the total skin and flesh counts of Cape hake during chilled storage of wet fish and thawed frozen fish under different conditions (Simmonds and Lamprecht, 1980*b*,*e*). In the tests from which these data are drawn, thawing was performed as rapidly as possible in air at about 20°C and moving at about 250 m min^{-1}, the fish being protected from the air by thin polythene film. The fish were placed into the appropriate chill storage environment as soon as thawing was complete, as judged by the flexibility of the fish. In this way it was hoped that any extraneous effects due to the thawing mechanism would be minimised.

In all cases spoilage, as assessed by odour, was slower in thawed frozen fish than in wet fish. The growth rate of bacteria in thawed fish

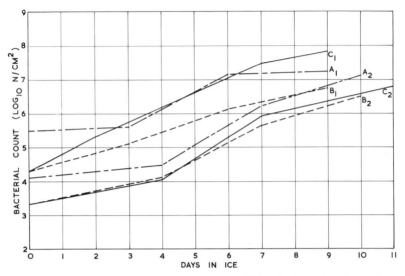

Fig. 5. Increase of total skin counts at 20°C of three batches of wet (A_1, B_1, C_1) and thawed shore-frozen (A_2, B_2, C_2) hake during storage in ice.

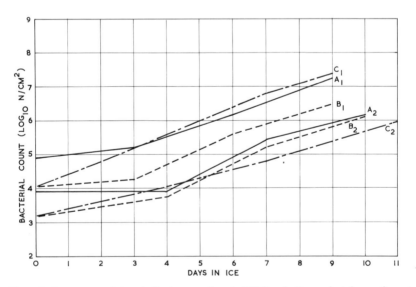

Fig. 6. Increase of total flesh counts at 20°C of three batches of wet (A_1, B_1, C_1) and thawed shore-frozen (A_2, B_2, C_2) hake.

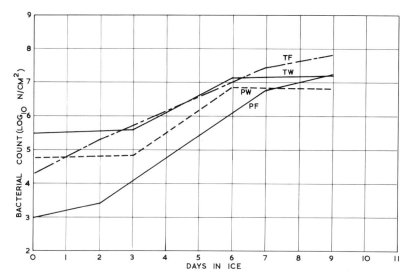

Fig. 7. Increase in 20°C skin counts of wet and thawed shore frozen hake during storage in ice. TF = total count, thawed frozen; TW = total count, wet; PF = *Pseudomonas* count, thawed frozen; PW = *Pseudomonas* count, wet.

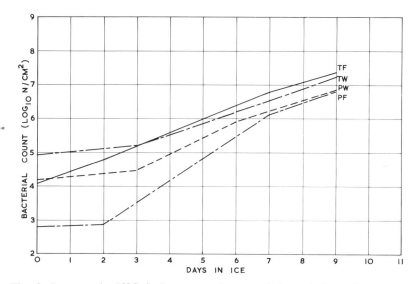

Fig. 8. Increase in 20°C flesh counts of wet and thawed shore frozen hake during storage in ice. TF = total count, thawed frozen; TW = total count, wet; PG = *Pseudomonas* count, thawed frozen; PW = *Pseudomonas* count, wet.

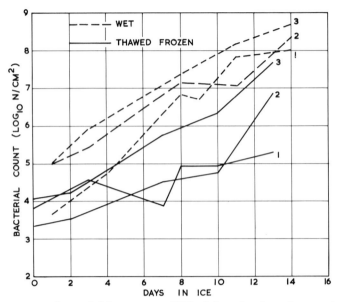

Fig. 9. Increase in total skin counts at 20°C of three batches of wet and thawed sea frozen hake during storage in ice.

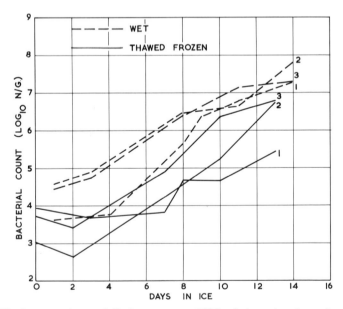

Fig. 10. Increase in total flesh counts at 20°C of three batches of wet and thawed sea frozen hake during storage in ice.

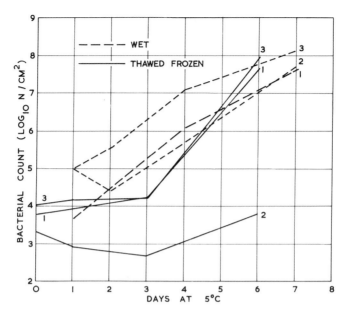

Fig. 11. Increase in total skin counts at 20°C of three batches of wet and thawed sea frozen hake during storage in air at 5°C.

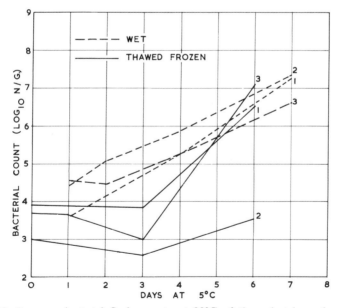

Fig. 12. Increase in total flesh counts at 20°C of three batches of wet and thawed sea frozen hake during storage in air at 5°C.

was slower, and the lag phases more extended. These effects were more pronounced for sea frozen fish than for shore frozen, and for fish held at 5°C than for equivalent ones stored in ice.

Figures 5 and 6 demonstrate that, in all cases, the total count over the normal storage range of thawed frozen hake was lower than that for the equivalent fish stored under the same conditions in ice without having been frozen. Figures 7 and 8 show data for two batches selected from these lines. It can be seen that, even where the total count of the thawed fish exceeds that for the wet fish, the total *Pseudomonas* count of thawed fish generally remains lower. Figures 9 and 10 show similar, but somewhat more pronounced, effects for sea frozen fish stored in ice. Finally, in Figs 11 and 12, under the accelerated spoilage rates obtaining at 5°C, it can be seen how the growth rates on thawed fish can, after an extended lag phase, be sufficiently great to result eventually in a higher count for thawed frozen fish than for wet fish. Even under these conditions the quality of the former fish was acceptable, while the wet fish were past the limits of edibility (Simmonds and Lamprecht, 1980*b*).

Despite the real enhancement of storage life of frozen and thawed fish as compared with wet fish, it must be emphasised that these benefits are probably difficult to obtain in practice, particularly if strict control is not maintained over the thawing process. As it is safer to assume that freezing merely preserves the initial quality of the fish, so it is equally more prudent to consider the thawed material no better than before freezing, from a bacteriological or any other standpoint.

THE EFFECT OF HANDLING AND PROCESSING

The effect of handling and processing on the microbial population of fish is twofold—the numbers of existing bacteria may increase by incubation, and new organisms may be introduced by contamination. The former may be kept to a minimum by maintaining temperatures at all stages as low as possible, consistent with the comfort of the worker; the latter, by regular and frequent cleaning and disinfecting of all the potential sources of infection.

The first source of contamination, and perhaps the most difficult to eradicate, is the fishing vessel itself. The surfaces of the deck, pound boards, storage tanks and holds are all breeding grounds for micro-organisms, and unless cleaned at every opportunity between operations, can cause needless shortening of the storage life of the fish.

The catching method can have an effect on the rate of spoilage. Trawled fish generally do not keep as long as line-caught fish, firstly because the pressure in the net expresses the contents of the digestive tract and spreads them over the skin surfaces of the catch, and secondly because line-caught fish, having struggled less before death, tend to remain in rigor mortis longer. While ice direct from the ice plant usually has low counts with high proportions of coryneforms and flavobacteria, it rapidly picks up fish spoilage bacteria, largely pseudomonads, by contact with the hold, pound boards and the shovels used to distribute the ice (Georgala, 1957). Impervious surfaces such as metal can have just as high a count as wooden ones; they are, however, more easily cleaned and disinfected.

In many countries it is the practice to auction fish on the quayside or in fish markets. The fish are often presented in open boxes, without icing, and considerable increases in counts can arise from prolonged holding at elevated temperatures. Further contamination can also occur not only from contact with boxes and floors but from airborne sources.

It might be thought that since the primary sources of bacteria in freshly-caught fish are the skin, gills and intestines, a considerable reduction in count could be obtained in the final product by heading, gutting, filleting and skinning the fish as soon as possible. However, cross-contamination of the cut surfaces by knives, hands and machinery can occur, despite the most diligent use of antiseptic dips, chlorinated processing water and general cleanliness. Indeed, without such precautions, the bacterial loads would generally be greatly increased by processing.

Similar considerations apply in many instances to crustaceans. Although the head of shrimp is about 40% of the total body weight, it contains about 75% of the bacterial load (Novak, 1973) and beheading can reduce counts by 50% and more. Other evidence indicates that little difference exists from a microbiological point of view between the flesh of whole and beheaded shrimp, and that beheading at the time of catching is more likely to be activated by economic and technological considerations (Cann, 1977).

In recent years, increasing amounts of fish have been frozen at sea, and then thawed ashore for further processing. It is often the practice, for reasons of expediency, to thaw large batches at once, for instance sufficient material for a whole day's processing. In such cases, it is imperative that an adequate chill storage facility is available for holding the thawed fish prior to processing. If this is not done, not only will there be a rapid growth of spoilage bacteria, but any mesophilic organisms introduced by handling will also be able to increase, with the attendant possibility of development of toxic substances. A similar situation can arise during cold smoking. In modern processing, the purpose of smoking is merely to introduce a desirable flavour, and the bactericidal property of the smoke under these conditions is negligible. Often the drying effect is deliberately kept to a minimum for reasons of increased yield, and the net result is an increase in count, again with the possibility of development of pathogens.

Changes in the numbers and types of bacteria are also brought about (before smoking) by the brining process (Shewan, 1961). The loads in the brining tank may increase a hundredfold during a production run, with a corresponding increase in the contamination of the immersed fish. Hot smoking, on the other hand, can reduce the total count virtually to zero, but unless the product is rapidly chilled and frozen as soon as possible, the benefits obtained will be lost (Simmonds *et al.*, 1974, 1975).

MINCE AS A RAW MATERIAL

Fish mince obtained from deboning machines is increasingly used as more efficient use of raw materials assumes greater importance, and its properties render it especially prone to rapid bacterial spoilage (Babbitt, 1972; Babbitt *et al.*, 1974; Waters and Garrett, 1974; Cann and Taylor, 1976; Crabb and Griffiths, 1976; Licciardello, 1980). The process of mincing destroys the structure of the flesh, increases the surface area, makes nutrients more readily available to micro-organisms and spreads the bacterial load through the mince. In addition, the quality of the water used in processing is of even greater importance than usual, as it gets into more intimate contact with the material.

Provided that adequate hygiene is practised, the initial bacteriological quality of mince will depend on that of the starting product. Care should be taken to ensure that only clean materials are used and that slime, viscera, blood and foreign matter are excluded. Higher counts are usually obtained with frame mince than with that from fillets or whole fish. The mince should be kept chilled while awaiting incorporation into other products, such as fish cakes, blocks and re-formed products. At chill temperatures, bacterial growth is essentially aerobic and takes place mainly on the surface, being relatively slow in the middle. It would appear safe to store mince for 24 hours at temperatures of 5°C or lower.

It has been suggested that bacteriological standards specifying total counts of less than $10^5 \, g^{-1}$ are too severe (Hobbs *et al.*, 1971), and in a survey of 28 Canadian plants, Blackwood (1974) found that only 60% of samples had total plate counts at 25°C of less than $10^6 \, g^{-1}$. Only the freshest iced fish, say less than four days old, can be expected to provide mince which meets the bacteriological standards proposed. Mincing should not, therefore, be used to salvage poor quality fish. It is worthy of note that, in the Japanese industry, minced fish is produced under stringently controlled conditions of sanitation, resulting in a material of highly satisfactory bacterial quality for use in fish sausage. It should equally be realised, however, that unless the principles of sanitation are rigidly adhered to, this product is extremely prone to bacterial contamination.

PREPARED FROZEN FISH PRODUCTS

The frozen food industry has expanded and diversified greatly in recent years, and a large range of raw and pre-cooked frozen products is now available. These range from such simple products as battered or breaded fillets, fish sticks and fish portions, to shaped and extruded materials incorporating starches and other ingredients and additives, and finally to convenience foods comprising fish, sauce and pasta, potato or other vegetables.

As in raw fish, operations such as filleting or sawing-up frozen blocks to produce fingers or portions, can introduce contaminants via accumulation of sawdust on blades and handling by operators. A further source of micro-organisms could be the ingredients, such as starches, breadcrumbs, herbs and spices, egg and dairy products, which may be used in such prepared foods. For example, the frequent use of egg in batter formulations has given rise to concern over the possibility of introducing *Salmonella*. In a study of a processed seafood plant, it was found that relatively low levels of bacteria of public health significance entered the plant via the raw fish material (Raj and Liston, 1963). The initial cutting caused a tenfold increase in counts and also introduced coliforms, enterococci and *Staphylococcus aureus*. Further contamination was introduced by battering and breading. Larkin *et al.* (1956) suggested that the breading procedure was a major source of contamination. Pre-cooking, such as frying to set the coating, does not necessarily reduce the bacterial load significantly (Raj, 1970).

Since many micro-organisms survive freezing and cold storage, it is necessary to determine whether the conditions of cooking, reheating and other handling, in the home or in catering establishments, are adequate to ensure safety to the consumer. Heating procedures recommended by the manufacturers of frozen foods do not always result in high enough temperatures to destroy pathogenic bacteria. Houghtly and Liston (1965) showed that *S. aureus* was killed only if the food was heated to at least 49°C for 5 min, and that even then, viable organisms could reappear as a result of regrowth of the few survivors. This temperature could be attained in frozen fish sticks by heating in a hot air oven at 204°C for 12 to 15 min.

In another study, it was found necessary to heat frozen prepared fish dishes for at least 35 min in hot air at 200°C, or 12 min in a microwave oven of 600 W at maximum setting, in order to ensure a minimum temperature of 60°C at the centre (Atkinson *et al.*, 1980). Results were naturally variable as the slab thickness ranged from 2·75 to 3·60 cm, but edge temperatures after these treatments were of the order of 80–85°C. Coliforms and staphylococci were largely destroyed by cooking, but sporeformers readily survived normal cooking treatments. For fish cakes and portions, deep-frying gave better results than other cooking methods, acceptable temperatures being attained in 4–6 min at 180°C, compared with 25 min at 190°C in a hot air oven, and organoleptic quality was distinctly superior.

It is imperative, therefore, that the cooking method be tailored to the size, packaging and type of product, and be compatible with an acceptable organoleptic quality. Within these restrictions, the longest possible cooking time should be employed, preferably to attain a minimum internal temperature of 70–75°C.

AVOIDING PROBLEMS WITH FROZEN FISH PRODUCTS

The same principles of sanitation and hygiene which apply to the handling and processing of all foodstuffs apply equally to the production of frozen fish products, from the original catching operation on board the fishing vessel up to cold storage and distribution. At all stages scrupulous cleanliness must be observed, regular disinfection applied and temperatures maintained as low as possible consistent with the particular phase of processing.

Three classes of disinfectant for fish processing areas and equipment are regularly and effectively employed, viz. chlorine or chlorine-based compounds, iodophors and quaternary ammonium compounds (QACs). Chlorine-releasing compounds are inactivated by the presence of organic matter, and consequently require prior cleaning of dirty surfaces preferably with a detergent. They are, however, inexpensive and convenient to apply to fish handling machinery, particularly on board. They have been found to be more effective against spores than iodine-based compounds (Cousins and Allan, 1967). Iodophors and QACs are formulated to include surface active agents, and they have been found to be more effective than chlorine for disinfecting surfaces contaminated with fish material (Lamprecht and Pery, 1979). Iodophors have been used to treat shellfish depuration water (Fleet, 1978), and to destroy *Vibrio parahaemolyticus* contamination in oysters (Gray and Hsaio-Lin Hsu, 1979). Care must be taken when water containing halogens, particularly chlorine, comes into direct contact with fish being processed, because reactions between halogens and organic compounds can give rise to severe loss of flavour, or even 'medicinal' taints in the fish thus treated (Atkinson, 1965; Nachenius and Atkinson, 1973; Simmonds and Coetzee, 1974). The tainting effect is increased if stale fish or fish material is involved, and minced fish is particularly liable to tainting owing to the contact of the water with the fish material. Recirculation of process water should be avoided at all stages, as dirty water requires the addition of large amounts of chlorine to obtain a residual of even a few parts per million. All surfaces that come into contact with fish must be water and corrosion resistant, and easy to clean. Where wooden construction is employed, as is often the case on board fishing vessels, the wood should be sealed and coated with an impervious non-toxic material, such as a suitable paint. On-board cleaning is more difficult than in installations ashore owing to limited space, and high pressure jets of sea water are often the most effective means available. Fish holds and storage tanks must be thoroughly cleaned and disinfected between voyages and left-over ice discarded. It is essential to clean and disinfect fish working areas and machinery after completion of handling each catch. Working areas, processing machinery and holds should also be designed to facilitate cleaning; holds in particular must be well drained.

If adequate chilling cannot be applied to the catch immediately, the fish must at least be protected from exposure to direct sunlight and wind, and should be handled and processed in a sheltered area. Immediate stunning of line-caught fish will prevent quality loss caused by struggling and exhaustion. Gutting and bleeding should be carried out as soon as possible after catching. Gutting must be complete; if not, it may cause more contamination than no gutting at all. Guts should be removed from the fish handling area immediately and the fish thoroughly washed. If roes or livers are kept, separate storage should be provided for them.

Fishing voyages should not be longer than is consistent with maintaining acceptable quality. If chilling facilities are not available, catches must be transferred ashore or to a factory vessel within a few hours, or at most, one day. When fish are held chilled for freezing elsewhere, they should be discharged while still fresh enough to produce an acceptable frozen product after the normal delays and temperatures attendant on processing for freezing. The voyage should, in fact, be short enough to allow for a comfortable margin of safety; fish must never reach the freezer in a barely edible condition.

Icing should be carried out in such a way as to ensure that all fish are in intimate contact with the ice. If chilled or refrigerated sea water is used, the capacity of the system should be sufficient to ensure that temperatures of $-1°C$ to $0°C$ are obtained. For extended storage, sea water may be gradually replaced to remove contaminants without increasing the temperature.

Lobsters and crabs should be kept alive until processing and freezing. If the fishing grounds are too distant to achieve these objectives, the only acceptable course is to freeze at sea.

Sea frozen fish should be transferred to freezer installations ashore as rapidly as possible to avoid undesirable rises in temperature. Wet fish must be stored in a chill room to await further processing and, better still, be iced. Deterioration is more rapid in air, even at $0°C$, than in ice, and furthermore ice will more rapidly restore the desired low temperature if fish have warmed up during off-loading and transport to the factory. Fish and other ingredients must be examined immediately on arrival at the factory, and, if rejected, removed immediately from the processing area and disposed of, as soon as possible, to avoid contamination. Likewise, filleting and other waste, especially viscera, heads, and shells of crustaceans, should not be allowed to accumulate on the processing line. Unused batter or breadcrumbs should be disposed of before an excessive build-up of bacterial load occurs.

Delays of more than a few minutes during filleting and processing should be avoided, and if this is not possible, the fillets or other raw materials should be chilled, preferably with ice, while awaiting further handling. Hot, cooked ingredients or products should be cooled as rapidly as possible, at the appropriate stage of processing, to prevent incubation of any residual micro-organisms. Products should be packed as soon as they are ready, and packages should not lie around awaiting freezing. Freezer trolleys should be kept in a cool area, preferably at 10°C or lower, while being loaded, and transferred to the freezer as soon as they are filled. The product should remain in the freezer until completely frozen, and then be transferred directly to the storage room without delay. When frozen products are removed from the cold store for distribution, there must be no delay before loading on board the transporting vehicle. The staging and loading areas should be kept as cool as possible, and the means of transportation should be capable of maintaining a suitably low temperature, never higher than −18°C.

When fish is thawed for further processing and re-freezing, thawing should be as rapid as possible consistent with maintaining acceptably low surface temperatures. Thawing media should not exceed 18°C (for still air) or 21°C (for water). At no stage during processing should raw fish be exposed to temperatures above about 20°C. Only potable fresh water, or sea water of equivalent microbiological quality, should be allowed to come into contact with the fish, and its temperature should be as low as possible, preferably not in excess of 10°C—except in the case of thawing water.

Factory premises should be constructed according to accepted principles of food processing hygiene, and should, together with all fish handling, conveying, processing, freezing and storage equipment, be designed for ease of cleaning, corrosion resistance, and freedom from features which encourage build-up of contamination. The surrounding grounds should be free from areas of loose sand and other possible sources of infection, and should preferably be under paving. In some fisheries, it is still difficult to avoid landing catches on beaches and similar areas, but such practices should be eliminated wherever possible.

Cutting and filleting boards must be of suitable hygienic material— wood is not recommended. These should be cleaned and disinfected thoroughly and frequently during working, as also must knives. Tables, conveyors and machines should be cleaned and disinfected before and after each production shift, and during every stoppage in production. It is important to remove accumulated waste water from conveyors and machines. This is especially relevant in the case of materials such as mince, or cooked ingredients for prepared products, such as fish in sauce or fish cakes. It is also desirable to refrain from processing more than one type of product on the same line during the same day, for without thorough disinfection during changeover, cross contamination can readily occur.

Guidelines have been laid down in considerable detail for the production of high quality frozen fish products, for example by the Food and Agriculture Organization of the United Nations (FAO, 1982).

REFERENCES

Alford, J. A., Tobin, L. and McCleskey, C. S. (1942). *Food Research*, **7**, 353.
Atkinson, Alison (1965). South African Fishing Industry Research Institute Annual Report, 19, pp. 26–7.
Atkinson, Alison, Lamprecht, Engela, Evans, Desireé and Pery, Lydia (1980). South African Fishing Industry Research Institute Progress Report 212.
Babbitt, Jerry K. (1972). *Proceedings Oak Brook Seminar on Mechanical Recovery and Utilization of Fish Flesh* (Ed. Roy E. Martin), National Fisheries Institute, Washington, DC, pp. 49–65.
Babbitt, J. K., Crawford D. L. and Law, D. K. (1974). *Proceedings Second Technical Seminar on Mechanical Recovery and Utilization of Fish Flesh* (Ed. Roy E. Martin), National Fisheries Institute, Washington, DC, pp. 32–43.
Barrow, G. I. and Miller, D. C. (1969). *Lancet*, 2, 421.
Barrow, G. I. and Miller, D. C. (1972). *Lancet*, 1, 485.
Barrow, G. I. and Miller, D. C. (1976). *Microbiology in Agriculture, Fisheries and Food* (Eds F. A. Skinner and J. G. Carr), Academic Press, London, pp. 181–93.
Blackwood, C. M. (1974). *Fishery Products* (Ed. Rudolf Kreuzer), Fishing News (Books) Ltd, London, England, pp. 325–28.
Bligh, E. G. (1971). *Fish Inspections and Quality Control* (Ed. Rudolf Kreuzer), Fishing News (Books) Ltd, London, England, pp. 81–4.
Bramstedt, Fritz and Auerbach, Margarethe (1961). *Fish as Food*, Vol. 1 (Ed. Georg Borgstrom), Academic Press, London, pp. 613–38.
Brown, R. K. and McMeekin, T. A. (1977). *Food Technology in Australia*, **29**, 103–6.
Cann, D. C., Taylor, L. Y. and Hobbs, G. (1975). *Journal of Applied Bacteriology*, **39**, 331–6.
Cann, D. C. and Taylor, L. Y. (1976). *Proceedings of Conference on the Production and Utilization of Mechanically Recovered Fish Flesh (Minced Fish)* (Ed. James N. Keay), Torry Research Station, Aberdeen, pp. 39–45.
Cann, D. C. (1977). *Proceedings of Conference on Handling, Processing and Marketing Tropical Fish* (Eds Penelope Sutcliffe and J. Disney), Tropical Products Institute, London, pp. 377–94.

Cousins, C. M. and Allan, C. D. (1967). *Applied Bacteriology*, **30**, 168.
Crabb, W. E. and Griffiths, D. J. (1976). *Proceedings of Conference on the Production and Utilization of Mechanically Recovered Fish Flesh (Minced Fish)* (Ed. James N. Keay), Torry Research Station, Aberdeen, pp. 46–8.
Cuddeford, D. (1983). *Proceedings of the Institute of Food Science and Technology (United Kingdom)*, **16**(3), pp. 130–4.
Cutting, C. L. (1955). *Fish Saving*, Leonard Hill (Books) Ltd, London, pp. 294–310.
Disney, J. G., Cole, R. C. and Jones, N. R. (1974). *Fishery Products* (Ed. R. Kreuzer), Fishing News (Books) Ltd, Surrey, England, pp. 329–37.
Dyer, F. E. (1947). *Journal of the Fisheries Research Board of Canada*, **7**, 128–36.
Early, J. C. (1967). M.Sc. Thesis, University of Nottingham.
Fieger, Ernest A. and Novak, Arthur F. (1961). *Fish as Food*, Vol. 1 (Ed. Georg Borgstrom), Academic Press, London, pp. 561–611.
Fleet, G. (1978). *Food Technology in Australia*, **30**(11), 444–54.
Frazier, W. C. (1967). *Food Microbiology*, 2nd Edn, McGraw-Hill Book Co., p. 283.
FAO (1967), *Yearbook of Fishery Statistics*, FAO, Rome, 23, 1966, pp. XX–XXI.
FAO (1969). *Yearbook of Fishery Statistics*, FAO, Rome, 27, 1968, pp. a4–a5.
FAO (1979). *Yearbook of Fishery Statistics*, FAO, Rome, 47, 1978, pp. 2–3.
FAO (1982). Reference Manual to Codes of Practice for Fish and Fishery Products, Fisheries Circular No. C750, FAO, Rome, 257 pp.
FAO (1983). *Yearbook of Fishery Statistics*, FAO, Rome, 53, 1981, p. 20.
Graham, J. (1982). *Fish Handling and Processing* (Eds A. Aitken, I. M. Mackie, J. H. Merritt and M. L. Windsor), Torry Research Station, HMSO, Edinburgh, pp. 56–78.
Georgala, D. L. (1957). Ph.D. Thesis, Aberdeen University.
Georgala, D. L. (1958). South African Fishing Industry Research Institute Annual Report 12, p. 10.
Gray, R. H. and Hsaio-Lin Hsu, D. (1979). *Journal of Food Science*, **44**(4), 1097–100.
Heen, E. and Karsti, O. (1965). *Fish as Food*, Vol. IV (Ed. Georg Borgstrom), Academic Press, London, pp. 355–418.
Hess, Ernest (1934a). *Contributions to Canadian Biology and Fisheries*, **8**, 461–74.
Hess, Ernest (1934b). *Contributions to Canadian Biology and Fisheries*, **8**, 491–505.
Hobbs, G., Cann, D. C., Wilson, Barbara B. and Horsley, R. W. (1971). *Journal of Food Technology*, **6**, 233–51.
Houghtly, G. and Liston, J. (1965). *Food Technology*, **19**(5), 192–5.
Ienistea, C. (1973). *The Microbiological Safety of Food* (Eds Betty C. Hobbs and J. H. B. Christian), Academic Press, London, pp. 327–43.
Iyer, T. S. G., Choudhuri, D. R. and Pillai, V. K. (1971). FAO Fishery Reports 115, pp. 59–69.
Jay, J. M. (1978). *Modern Food Microbiology*, Second Edition, D. van Nostrand Company, New York, 479 pp.

Jones, N. R. (1965). *Fish Handling and Preservation*, Organisation for Economic Co-operation and Development, Paris, 31–56.
Jones, D. T. (1977). *Proceedings of the Conference on the Handling, Processing and Marketing of Tropical Fish* (Eds Penelope Sutcliffe and J. Disney), Tropical Products Institute, London, pp. 37–44.
Karthiayani, T. C. and Iyer, K. (1967). *Fish. Technol.*, **4**, 89–97.
Karthiayani, T. C. and Iyer, K. (1971). *Fish. Technol.*, **8**, 69–79.
Kawabata, Toshiharu (1971). *Proceedings SOS/70 Third International Congress Food Science and Technology*, Institute of Food Technologists, Chicago, pp. 692–6.
Kimata, Masao (1961). *Fish as Food*, Vol. I (Ed. Georg Borgstrom), Academic Press, London, pp. 329–52.
Kim, I. and Bjeldanes, L. F. (1979). *Journal of Food Science*, **44**, 922–3.
Kochanowski, J. and Maciejowska, M. (1964). *Prace Morskiego Instytutu Rybackiego*, **13** (Ser. B), 93–106.
Kochanowski, J. and Maciejowska, M. (1969). *Prace Morskiego Instytutu Rybackiego*, **14** (Ser. B), 153–67.
Lamprecht, Engela C. (1980). *Journal of the Science of Food and Agriculture*, **31**, 1309–12.
Lamprecht, Engela and Elliott, Marguerite (1971). South African Fishing Industry Research Institute Annual Report 25, p. 78.
Lamprecht, Engela and Ferry, Karen (1981). South African Fishing Industry Research Institute Annual Report 35, p. 70.
Lamprecht, Engela and Pery, Lydia (1979). South African Fishing Industry Research Institute Progress Report 200.
Larkin, E. P., Litsky, W. and Fuller, J. E. (1956). *American Journal of Public Health*, **46**, 464.
Laycock, R. A. and Regier, L. W. (1970). *Applied Microbiology*, **20**, 333–71.
Lee, J. S. and Harrison, J. M. (1968). *Applied Microbiology*, **16**, 1937–8.
Lee, J. S. and Pfeifer, D. K. (1975). *Applied Microbiology*, **30**, 72–8.
Lekshmy, A., Choudhouri, D. R. and Pillai, V. K. (1969). *Freezing and Irradiation of Fish* (Ed. Rudolf Kreuzer), Fishing News (Books) Ltd, London, pp. 343–7.
Lerke, P., Adams, R. and Farber, L. (1965*a*). *Applied Microbiology*, **11**, 458–62.
Lerke, P., Adams, R. and Farber, L. (1965*b*). *Applied Microbiology*, **13**, 625–30.
Licciardello, Joseph J. (1980). *Proceedings Third National Technical Seminar on Mechanical Recovery and Utilization of Fish Flesh* (Ed. Roy E. Martin), National Fisheries Institute, Washington, DC, pp. 458–76.
Liston, J. (1955). Ph.D. Thesis, Aberdeen University.
Liston, J. (1956). *Journal of General Microbiology*, **15**, 305–14.
Liston, J. (1957). *Journal of General Microbiology*, **16**, 205–16.
Liston, J. (1965). *The Technology of Fish Utilization* (Ed. R. Kreuzer), Fishing News (Books) Ltd, London, pp. 53–7.
Liston, J. (1980). *Advances in Fish Science and Technology* (Ed. J. J. Connell), Fishing News (Books) Ltd, London, pp. 138–57.
Mathen, C., Lekshmy, A., Pillai, V. K. and Bose, A. N. (1965). *The*

Technology of Fish Utilization (Ed. R. Kreuzer), Fishing News (Books) Ltd, London, pp. 206–9.

Nachenius, R. J. and Atkinson, Alison (1973). South African Fishing Industry Research Institute Annual Report 27, p. 8.

Newell, B. S. (1973). CSIRO Division of Fisheries & Oceanography Technical Paper 35.

Novak, A. F. (1973). *Microbial Safety of Fishery Products* (Eds C. O. Chichester and H. D. Graham), Academic Press, London, pp. 59–73.

Omura, J., Price, R. J. and Olcott, H. S. (1978). *Journal of Food Science*, **43**, 1779–81.

Raj, H. D. (1970). *Laboratory Practice*, **19**(4), 374–8.

Raj, H. D. and Liston, J. (1961). *Food Technology*, **15**, 429–34.

Raj, H. D. and Liston, J. (1963). *Food Technology*, **17**, 83–9.

Ray, S. M. (1971). *Proceedings SOS/70 Third International Congress Food Science and Technology*, Institute of Food Technologists, Chicago, pp. 717–28.

Reay, G. A. and Shewan, J. M. (1949). *Advances in Food Research*, **2**, 343–98.

Rowan, A. N. (1956). South African Fishing Industry Research Institute Annual Report 10, pp. 8–11.

Scholes, Robina B. and Shewan, J. M. (1964). *Advances in Marine Biology*, **2**, 133–69.

Shewan, J. M. (1944). *Proceedings of the Society of Agricultural Bacteriologists* (Abstracts) 1–5.

Shewan, J. M. (1949). *Journal of the Royal Sanitary Institute*, **69**(4), 394–421.

Shewan, J. M. (1961). *Fish as Food*, Vol. I (Ed. Georg Borgstrom), Academic Press, London, pp. 487–560.

Shewan, J. M. (1962). *Recent Advances in Food Science* (Eds J. Hawthorne and J. Muile Leitch), Butterworth, London, pp. 167–93.

Shewan, J. M. (1977). *Proceedings of Conference on Handling, Processing and Marketing Tropical Fish* (Eds Penelope Sutcliffe and J. Disney), Tropical Products Institute, London, pp. 51–66.

Simmonds, C. K. and Coetzee, P. (1974). South African Fishing Industry Research Institute Annual Report 28, pp. 12–13.

Simmonds, C. K. and Lamprecht, E. C. (1980*a*). *Advances in Fish Science and Technology* (Ed. J. J. Connell), Fishing News (Books) Ltd, London, pp. 297–9.

Simmonds, C. K. and Lamprecht, E. C. (1980*b*). *Advances in Fish Science and Technology* (Ed. J. J. Connell), Fishing News (Books) Ltd, London, pp. 417–21.

Simmonds, C. K. and Lamprecht, Engela (1980*c*). South African Fishing Industry Research Institute Annual Report 34, pp. 86–8.

Simmonds, C. K. and Lamprecht, Engela (1980*d*). South African Fishing Industry Research Institute Annual Report 34, pp. 88–91.

Simmonds, C. K. and Lamprecht, Engela (1980*e*). Unpublished results.

Simmonds, C. K. and Lamprecht, Engela (1981). South African Fishing Industry Research Institute Annual Report 35, pp. 72–3.

Simmonds, C. K. and Lamprecht, Engela (1982). South African Fishing Industry Research Institute Annual Report 36, 47–9.

Simmonds, C. K., Lamprecht, Engela, Coetzee, P. and Gouly de Chaville, Charmaine (1974). South African Fishing Industry Research Institute Annual Report 28, pp. 78–9.

Simmonds, C. K., Lamprecht, Engela, Rutkowski, H. P. M. and Gouly de Chaville, Charmaine (1975). South African Fishing Industry Research Annual Report 29, pp. 45–6.

Sreenivasan, A. (1959). *Journal of Scientific & Industrial Research (New Delhi)*, **18**c, 119.

Tarr, H. L. A. (1942). *Journal of the Fisheries Research Board of Canada*, **6**, 74–89.

Thjøtta, Th. and Sømme, O. M. (1938). *Acta Pathologica et Microbiologica Scandinavia*, (Supplement) **37**, 514–26.

Venkataraman, R. and Sreenivasan, A. (1955). *Current Science (India)*, **24**, 380–1.

Walker, P., Cann, D. and Shewan, J. M. (1970). *Journal of Food Technology*, **5**, 375–85.

Waters, Melvin, E. and Garrett, E. Spencer III (1974). *Second Technical Seminar on Mechanical Recovery and Utilization of Fish Flesh*, (Ed. Roy E. Martin), National Fisheries Institute, Washington, DC, pp. 217–36.

Williams, O. B., Campbell, L. L. and Rees, H. B. (1952). *Texas Journal of Science*, **4**, 53.

Wood, E. J. F. (1953). *Australian Journal of Marine and Freshwater Research*, **4**, 160–200.

Yoshimizu, M. and Kimura, T. (1976). *Fish Pathology*, **10**, 243–59.

Zobell, C. E. (1946). *Marine Microbiology*, Chronica Botanica, Waltham, Massachusetts, 240 pp.

Chapter 6

Microbiology of Frozen Dairy Products

J. Rothwell

Department of Food Science, University of Reading, UK

INTRODUCTION

The frozen dairy product which probably comes to mind first is ice cream, but many other dairy products are now frozen for one purpose or another, some on a very large scale indeed. In almost all of these cases, the intention is to increase the storage life of the product by a considerable margin, in order to improve its availability at times of the year when, because of seasonal production, it may not normally be so.

Among the other products which are regularly frozen and kept in a deep frozen state (at temperatures of −20°C or lower) are: liquid milk; concentrated milks; cream of various fat contents from about 9 per cent to 80 per cent; butter; cheese of many varieties, and cheese curd; starter cultures for most dairy purposes; and various dairy desserts.

The problems which arise in the freezing of some, or all, of these products are largely concerned with changes in their physical state which occur during freezing and/or thawing. In particular, the fat globule membrane is very easily denatured, and this results in a product from which the fat separates on thawing, so leading to an oily layer on the surface of the product. It is also possible for protein destabilisation to occur, which gives a rough, gritty texture to the thawed product.

Once a product is frozen, there is very little change in its overall microbial population, apart probably from a slow reduction in the numbers of viable organisms. The microbial standard of frozen dairy products reflects, in general, the microbiological quality of the initial raw milk, the processing equipment and methods, and the packaging method and materials, for, provided that the frozen product is maintained at a satisfactory low temperature (e.g. $-20°C$ or lower), there should be no deterioration in the microbiological standard. There may, however, be physical and chemical changes which occur over a long period. These could affect the appearance, colour, texture and taste of the product, and may be the cause, eventually, of sufficient deterioration to make the product unacceptable.

MICROBIOLOGY OF RAW MILK

As the microbial standard and content of the final frozen dairy product is dependent to a very large extent on the quality of the raw materials, a review of the microbiology of these products must, logically, begin with the raw milk itself.

In almost all dairying areas in developed countries, freshly obtained milk is cooled on the farm, usually in a refrigerated bulk tank, to about 4°C. After inspection of the milk (appearance, smell and temperature), and sampling, the evening and morning milkings are collected by insulated road tanker. Along with the milk from other farms, it is pumped, on arrival at the receiving dairy, into a large refrigerated silo capable of holding 100 000 litres or more, and may be kept in the silo for 24 h, or in some cases even longer, at a temperature below 5°C.

The microbial content of the raw milk will, of course, be affected by the method of handling, and the organisms present will have come from one, or several, sources of contamination, including the interior and exterior of the udder, and the milking, cooling and storage equipment. Methods of assessment of the quality of cooled, raw milk have been reviewed in the International Dairy Federation (IDF) document 83 (IDF, 1974a), and the factors which influence the bacteriological quality in IDF document 120 (IDF, 1980).

Briefly, the microbial flora in the udder, will depend on whether the gland is healthy or infected. In a healthy udder, the milk may well be sterile, and with machine milking, only very low numbers of bacteria should be present if proper precautions are taken to avoid contamination; according to IDF document 120, the usual contaminating organisms are Micrococcaceae and *Corynebacterium bovis*. However, the teat canal may regularly contain various types and numbers of organisms, including staphylococci, corynebacteria, *Actinomyces*, coliforms, *Pseudomonas* spp., and *Proteus* spp. Infected cattle may excrete *Mycobacterium bovis, Brucella abortus, Salmonella* spp., *Listeria* and *Coxiella,* although in the UK cattle are now free of tuberculosis and brucella, and proper heat treatment (e.g. pasteurisation) should kill all pathogenic organisms and most other contaminating and spoilage organisms.

The most widespread disease of cattle is mastitis, which, although caused mainly by staphylococci and streptococci (pathogenic types), may also have *Corynebacterium, Pseudomonas aeruginosa* or Enterobacteriaceae (*Escherichia coli* for example) as the causative agent. All of these may be present in the milk, but they should be killed by adequate heat treatment during subsequent processing.

Contamination from the udder surface will obviously depend on the state of cleanliness of that surface, and even on whether the cattle are housed inside or outside; for example, fewer anaerobic spores may be found in milk if the cattle are grazing out of doors. The type of bedding and the effectiveness of udder washing will also play a part, and IDF document 120 mentions that the following organisms may be found: staphylococci, *Corynebacterium,* streptococci, *Bacillus, Actinomyces,* coliforms, *Pseudomonas* and *Proteus.*

Depending on the standard of cleaning and disinfection, the milking machine will also contribute to this microbial flora, and micrococci, *Corynebacterium* and streptococci have been shown to predominate (IDF document 120). It is also reported that there may be many psychrotrophic organisms present, but that their numbers are very variable.

The subsequent handling of the milk also contributes to the microbial flora, and initially, it will normally be kept in storage tanks at a temperature of 4°C to 5°C. There is, at present, a tendency for the raw milk to be stored at these relatively low temperatures for increasing lengths of time, and it is possible for the period between milking and the commencement of processing to be as long as 48 h.

Several psychrotrophic species of bacterium produce proteinases and lipases, and these enzymes are, in many cases, very heat resistant. Their presence has been reported by several workers (e.g. IDF 1974*b*; Muir *et al.,* 1979; IDF, 1981), and although organisms (such as *Pseudomonas*) will be killed by heat treatment, the enzymes remain active; the spoilage of UHT products is often the result of enzyme activity. However, during the frozen storage of milk and milk products, little or no change would be expected, even if enzymes were present, but trouble could arise when the product is thawing out and stands, awaiting use, at temperatures approaching ambient (e.g. around 20°C). Other, but similar, problems may arise if there is a relatively high population of *Bacillus* in the raw milk. In this case, a small number will possibly survive UHT treatment (with larger numbers surviving ordinary pasteurisation), and may cause spoilage by their enzymes; proteolytic and lipolytic types are both present (IDF, 1981).

Thus it will be seen that, although efficient heat treatment of milk before it is processed will reduce the bacterial numbers very considerably, there still remains the possibility of survival of some spores, and also of some heat resistant enzymes. Both of these elements may affect the final, thawed product, and, to reduce the possibility of faults caused by these agents, it is necessary not only to have raw milk of very high quality but also to begin the processing with as little delay as possible.

FROZEN MILK

Work on the freezing of milk has been carried out since about 1934, but it was only during and after the 1939–45 War that serious research began, initially with attempts to supply US troops with milk kept at −23°C. It was found that if fluid milk (not concentrated) was pasteurised, homogenised, cooled, filled into waxed cartons and then frozen, the product had a life of about 6 months, and often was acceptable for up to 9 months (Samuelsson *et al.*, 1957).

During the years of research which followed, the basic problems in the manufacture and storage of frozen milk were identified as follows:

1. Destabilisation of the fat emulsion.
2. Precipitation or flocculation of the milk protein.
3. Development of flavour defects.
4. Microbiological quality of the product after thawing.

Destabilisation of Fat Emulsion

In milk, the fat is maintained in an emulsified state by the fat globule membrane which consists of phospholipids and proteins. During freezing, fat globules are pushed together to form aggregates, producing pressure which, with continued freezing, leads to disruption of the membranes. Liquid fat may be extruded, and, on thawing, the fat globules then coalesce.

Attempts to overcome this problem have included homogenisation of the milk to reduce the size of fat globules and so obtain as fine a dispersion as possible (Babcock *et al.*, 1946), and the use of ultrasonic vibrations (Wearmouth, 1957); since then, little has appeared in the literature on this latter method.

Protein Destabilisation

Flocculation or coagulation of the milk proteins, particularly the casein as calcium caseinate, is also a major problem, especially during defrosting, or if the temperature of storage fluctuates. It appears that rapid freezing, and maintaining the temperature of the frozen milk at between $-20°C$ and $-40°C$, as steady as possible, will only lead to protein destabilisation if storage is prolonged (Babcock *et al.*, 1947). To preserve the physical properties, it is necessary to pasteurise and homogenise the milk before freezing, so that the bacterial population should be low, and without pathogenic organisms.

Flavour Defects

Flavour defects, such as stale, or flat tastes, and finally oxidisation, have been shown to develop, but again the incidence of off-flavours is reduced if the initial milk is very fresh, subjected to satisfactory heat treatment, and the frozen product is maintained at a low steady temperature ($-20°C$ to $-40°C$). Both flavour defects and protein coagulation may be reduced by the addition of sodium citrate ($2 \, \mathrm{g \, litre^{-1}}$) (Babcock *et al.*, 1951) and ascorbic acid (Bell, 1948).

The practicality of freezing milk on a large scale, however, is an economic one, and newer products have probably reduced the value of the process almost to nil; such products include UHT milk and dried milks. In the shorter term, there is probably a use for the freezing method for storing samples prior to chemical and microbial investigations.

However, there is an increasing demand for goats' milk, and this is very easy to freeze. In fact, goats' milk and sheep's milk may be frozen without any initial treatment (unless it is felt desirable to pasteurise the milk), merely by cartoning it and keeping it in a deep freeze cabinet at

−20°C. Such milk appears to undergo little or no change even after several months' storage. In addition, there is a small demand for human milk which can also be kept frozen.

Microbiological Aspects of Frozen Milk

The first, and very important, factor which affects the microbiological quality of frozen milk is the quality of the raw milk, but a close second is the efficiency of the heat treatment and the subsequent handling of the product.

As with all frozen products, provided that the temperature is kept steady and at a sufficiently low figure, there will be no microbial growth during storage, but there is considerable discrepancy amongst the available reports as to whether there is a significant reduction in numbers during this time. Some of the earlier work, in particular, gives figures showing a reduction in cell numbers between the beginning of the process and the final thawed product, after storage (at −30°C) for up to 115 days (e.g. Randell, 1949). There is, however, no general concensus on this point, and some investigators even report increases in counts; for example Giroux *et al.* (1971) found up to ten-fold increases in numbers of micrococci and staphylococci in raw milk stored for over five months in liquid nitrogen. Increases were also found in lactobacilli, coliforms, corynebacteria and *Pseudomonas,* but not on so great a scale.

Murray and Coey (1959), however, found no change in the bacterial content of milk stored at −10°C for up to four months, but after that, a marked fall occurred. When coliform organisms were added to the milk before it was frozen, similar changes occurred.

In work carried out to establish the effect of freezing on samples of milk from individual udder quarters being investigated for mastitis, Bashandy and Heider (1979) showed that, when the milk was kept at −80°C, similar numbers of staphylococci were isolated from fresh and frozen samples. The numbers of streptococci were slightly reduced in the frozen milk, with considerable reductions after 14-days' storage. Coliform organisms, however, appeared not to be affected by freezing. This work was largely confirmed by Storper *et al.* (1982), who stored raw milk samples at −18°C, and showed that reductions in viability of isolates did occur. In some cases very small reductions were recorded, but in some species, considerable losses occurred, e.g. after 14 days,

Staphylococcus aureus was reduced by 1·6%, *Escherichia coli* by 9·7%, *Streptococcus agalactiae* by 12·5%, and non-agalactiae streptococci by 30·8%. Further reductions of between 5 and 20% were recorded after another two weeks storage, and it would appear that the majority of organisms remain viable for up to 14-days' frozen storage in raw milk.

Coghill and Juffs (1978) confirm the feasibility of freezing, for short periods, raw milk samples required for subsequent bacteriological analysis. They found that there was little reduction in cell viability in milk frozen at −20°C for 24 h, but, after longer frozen storage, counts became less reliable. The addition of glycerol did not protect bacteria during storage.

The reported numbers of organisms surviving frozen storage in milk vary considerably, and two factors which may be responsible for this variation have been reported in the literature. One is the effect of fast or slow thawing (Gebre-Egziabher, 1982). This worker found that the destruction of *Escherichia coli*, *Pseudomonas aeruginosa* and *Saccharomyces cerevisiae* was significantly greater when frozen skim-milk was thawed slowly rather than quickly. In addition, he found that the recovery of viable organisms, by plating, was slightly higher when peptone water was used as a diluent, but the differences were not significant statistically.

The other factor, reported by Yano and Morich (1971), is the possibility that the sub-lethal injury of bacteria during the process can lead to lower recovery figures on certain media. Thus, when *Escherichia/Aerobacter* organisms were inoculated into sterilised skim-milk which was frozen and then stored at −20°C, up to 73 per cent of the viable population was found to be unable to grow on desoxycholate agar or violet red bile agar, although the injured cells showed little sensitivity to brilliant green lactose bile broth.

One interesting report (McKinney *et al.*, 1973) showed that aflatoxin M_1 activity decreased slowly in frozen raw milk, but, in samples of curds and whey, it was still stable over periods of up to 4 months.

It would appear, therefore, that although there may be some reduction in total counts on thawing frozen milk, the magnitude of this loss varies with the organism, the method of freezing, storage and thawing, and with the length of time of storage. In addition, it appears to depend on the method of enumeration of the organisms in the final product. It must be stressed that pathogenic organisms, if present in the initial raw milk, will almost certainly survive relatively long periods of frozen storage.

FROZEN CREAM

There are three major reasons for freezing cream, namely:

1. To provide a storage system for cream during the season of high milk production, so that butter manufacture can continue during times of low production.
2. As a means of storing cream for ice cream manufacture, as cream frozen at high production times should be lower in price. This use for frozen cream is probably becoming much less, particularly as the manufacture of anhydrous milk fat increases.
3. For the production of frozen cream in retail packs, so that the product can then be thawed in small quantities, and used as a convenience food. This part of the frozen cream market is in considerable competition with UHT cream, and probably has an advantage in that properly processed frozen cream should not have any 'heat-treated' flavour, as is the case with some UHT creams.

Problems Associated with Frozen Cream

Provided that the cream to be frozen is of a high quality, the only problems likely to be encountered are:

(a) Oxidative rancidity, and
(b) 'Oiling-off' caused by a breakdown of the fat emulsion.

Oxidative rancidity can normally be kept to the minimum by the use of good quality, raw cream heat-treated at a temperature of 75°C or preferably higher, avoiding copper contamination in any form, and keeping the product at a temperature below −20°C.

'Oiling-off' can be minimised by efficient homogenisation of the cream during the heat treatment process, and by very rapid methods of freezing; the addition of sugar has also been shown to reduce the amount of 'oiling-off'.

Methods of Freezing Cream

The methods available for the freezing of cream vary depending on the use that is to be made of the final thawed product. For example, the cream frozen for buttermaking will not need to be frozen in any special way, as deterioration in the physical nature of the emulsion will not be detrimental to the final product, i.e. after the thawed cream has been churned into butter. Such cream, frozen in spring and early summer, will have a higher colour and Vitamin A content, so the butter eventually made from it should be superior in this way to butter made, for example, from winter-produced cream. Similarly, cream frozen for subsequent manufacture into ice cream will not need any special freezing techniques.

However, cream for the liquid market will have to be treated in a much more careful way, and the method of freezing adopted will depend, to some extent, on the fat content of the cream. In most cases, however, rapid freezing in a blast freezer or in small volumes in a plate freezer, or in small metal containers partly immersed in low temperature calcium chloride brine, is necessary to avoid the rupture of the fat globule membrane, which would lead to excessive 'oiling-off' when the cream thaws. Some methods used for the rapid freezing of cream of 35 per cent fat content ('whipping cream') to 48 per cent ('double cream') are discussed below.

On arrival, the milk is stored at a low temperature (4°C), before being separated at 29°C to 30°C. The cream is then batch pasteurised at 65°C for 30 min, or HTST pasteurised at about 80°C for 15 s, cooled to 15°C, and allowed to stand for about 10 h. Subsequently, it is filled into 5 fl. oz. (140 ml) or 10 fl. oz. (280 ml) plastic containers, which are sealed, and then frozen in a blast freezer. It is claimed that, by this process, faults like oily consistency and granular texture are avoided, and there is no apparent oxidation of the fat (Anon. 1979).

Another interesting method is described by Chase (1981). In this system, the heat-treated and cooled cream is fed into moulds on a revolving plate suspended over a brine bath at −30°C (e.g. a Gram ice cream lollie machine); the circular plate has 180 rows of 14 moulds. As soon as the moulds are filled with cream, a rake with spikes on it is lowered over each row, and one spike penetrates into each mould. The filled moulds are in the brine for about five minutes, by which time, the cream should be frozen. The moulds are then dipped into a hot brine bath to loosen the sticks of cream, which are then lifted out and put on an overhead conveyer. The sticks are automatically pulled off the rake and drop, twenty at a time, into a plastic sachet. The sachets are then heat-sealed and stored at temperatures around −20°C.

A third method of quick freezing of cream, known as 'Pellofreeze', has been described by Löndahl and Åström (1979). Cream is frozen on a stainless steel belt which is sprayed with a refrigerated solution of propylene glycol and water. In two to four minutes the cream becomes a solid, frozen slab, and, at the outlet, this slab is broken into pellets which may be easily handled for thawing. Bacterial tests of this method show a lowering in the plate count and number of coliforms after freezing.

Microbiological Aspects

The microbiology of liquid cream has been fully covered by Davis (1981). He indicates that, as already stated for frozen milk, the microbiology of frozen cream will be very similar to that for freshly produced, heat-treated cream, and although freezing may kill a few organisms, by mechanical damage caused by the formation of ice crystals, freezing itself cannot be used to reduce bacterial numbers. The relatively few references to the bacteriology of cream tend to confirm this statement, although varying results have been reported. For example, Samuellsson et al. (1957) quote workers who found a moderate increase in the bacterial count (from 1700 to 7000 cfu ml^{-1}) of pasteurised, homogenised cream, which was kept for six months at −20·5°C, while Kratochvil and Vedlich (1961) reported that, in cream which had been frozen and then stored at −16°C to −18°C for up to seven months, counts of coli aerogenes, mould and lipolytic organisms declined substantially during storage.

It would appear, therefore, that although there may be some decline in microbial population during frozen storage, frozen cream is a product whose final quality will be largely governed by the quality immediately before it was frozen.

BUTTER

The principles of buttermaking, i.e. churning from cream until the fat emulsion is inverted, have been covered in many textbooks (e.g. McDowall, 1953), and a brief summary is given by Murphy (1981) of both the traditional and the continuous buttermaking methods. Murphy stresses that very high bacteriological standards are required of the initial cream, and that it should be given a high heat treatment. In addition, he suggests that total counts of less than 1000 cfu ml^{-1}, with less than one coliform, yeast or mould, must be obtained if a high keeping quality butter is to result.

It has already been mentioned that the use of anhydrous milk fat (AMF) as a material for the manufacture of many milk products and other foods is increasing, and, to some extent, displacing the use of butter. This trend reflects the fact that AMF does not normally require to be kept at temperatures below its freezing point to ensure long-term storage. However, a considerable amount of butter is being made and kept in deep frozen storage, then released either direct to the markets, or for blending with other butters.

One major problem with the freezing of butter is that this does take an appreciable time. Murphy (1981) states that as butter is predominantly a fat product and has a low thermal conductivity, a box (25 kg) initially at 9°C will take in excess of 3 days to reach a temperature of −8°C in a cold store at −10°C. As a box will normally be surrounded by other boxes, the cooling rate will be even slower than this in normal practice. Another problem is that butter often contains salt, and this appreciably lowers the freezing point of the water in the butter; low temperatures of −20°C to −30°C are, therefore, commonly employed.

Butter of very good keeping quality will keep for up to 12 months at −25°C, while even poor keeping quality butter is reported by Schulz (1964) to keep for 3 to 6 months. For economy of operation, Dixon and Rochford (1973) suggest that a temperature of −12°C is a satisfactory compromise, as chemical, organoleptic and microbiological tests showed no significant differences when the butter was stored between −12°C and −23·5°C; storage at −6·5°C resulted in significantly greater deterioration.

Manu *et al.* (1972) found that coliform organisms decreased rapidly in butter which had been frozen at −25°C to −30°C, held for 36 to 40 h at that temperature, then stored at −15°C to −18°C. After 30 days, these organisms had disappeared in all but those samples with a high initial count. Yeasts and moulds, however, appear to be highly resistant, and only decreased to 10 per cent of initial values after up to 90 days; this report shows the importance of avoiding yeast and mould contamination during the manufacture of butter. This work is confirmed in a similar report published by Kastornykh *et al.* (1976), who, in addition, suggested that sorbic acid treatment of the lined, foil wrapper was useful in increasing storage life by the inhibition of mould growth.

Butter is, of course, a relatively poor medium for the growth of micro-organisms, particularly in the interior of the pack. This was illustrated by Sims *et al.* (1969) who contaminated unsalted and salted batches of butter with *Salmonella typhimurium*. In samples of salted butter stored at −18°C or −23°C, the mean survival rate was significantly reduced compared with storage at 0°C.

It is now usual practice to store butter at temperatures between −10°C and −18°C (Richards, 1982), for between these limits there appears to be a reduction in microbial counts. In addition, these low temperatures are not so excessive that they demand an uneconomic use of refrigeration.

FROZEN STARTER CULTURES

Starter cultures—cultures of harmless organisms used in the production of cheese and fermented milks—are required in large quantities by the dairy industry, and, indeed, in many other parts of the food industry as well.

Normally the starters are maintained in the form of 'mother cultures' in a sterile medium (usually skim or whole milk) under aseptic conditions, and from these mother cultures, a working culture is produced each day for inoculation into the bulk milk for the manufacture of the product. The mother cultures are usually prepared by the addition of a commercially manufactured, freeze-dried culture into a relatively small quantity (e.g. 500 ml) of sterile milk. These cultures are often kept in a frozen condition at the manufacturing creamery, to be thawed and used as necessary.

The whole process is lengthy, time consuming and may lead to contamination of the working culture unless great care is taken, and hence the use of freezing as a means of starter preservation is now becoming more widespread. It may be employed in one of three methods (Stadhouders *et al.*, 1969):

(a) the freezing of inoculated and incubated milks for 'mother cultures';
(b) the freezing of bulk starters; or
(c) freezing of concentrated starters.

Method (b) will be used if larger quantities of starter than just the mother culture are required, and bulk starters (i.e. not concentrated) may be frozen at −20°C to −40°C (Tamime, 1981), and then kept at that temperature for several months. Prolonged storage at −40°C can lead to a reduction in starter activity and may damage lactobacilli, but the use of specially formulated media can improve survival rates: one such medium, quoted by Tamime, contained 10 per cent skim milk, 5 per cent sucrose, fresh cream, and 0·9 per cent sodium chloride.

The freezing of concentrated starter is best carried out at around $-196°C$. Such concentrated starters may contain up to 10^{10} to 10^{12} cfu ml^{-1}, and may remain active for long periods in the presence of certain cryogenic materials, including sodium citrates, glycerol or sodium-B-glycerophosphate; cultures of both mesophilic and thermophilic starters have been so treated.

The starters are produced under carefully controlled conditions, and then concentrated (Gilliland, 1977), and the survival rate depends on the processing conditions and on the method of cell concentration. The usual methods of concentration involve either centrifugation, which can cause some damage to the cells, or continuous growth and neutralisation (at about pH 6) to produce high cell numbers. The limit, in the latter case, is imposed by the production of lactate, and some recent techniques (Osborne, 1977) enable the lactate to be removed from the medium by diffusion; cell concentrations of 10^{11} cfu ml^{-1} may be obtained in this way. The cell concentrates are then suspended in a small quantity of medium containing milk solids (Gilliland, 1977), although Stadhouders *et al.* (1969) reported that the addition of glycerol to the medium was important in protecting the bacteria during freezing. In addition to the total number of organisms surviving, it is essential that the activity of the starters should be maintained, particularly if mixed strain cultures are being frozen.

According to Gilliland (1977), there may be a close relationship between the amount of polysaccharide material present in the cells (expressed as cellular glucose) and the number of survivors, and in addition, the fatty acid composition of the cell membranes may also be important.

FROZEN CHEESE

Cheese is, of course, a milk product designed for long keeping, or, at the very least, with an extended shelf-life, and there has, therefore, not been much need for the development of methods for freezing cheese. Indeed, from the economic aspect, the long ripening time required for cheddar, and to a lesser extent other territorial cheeses made in the UK, is a major drawback in terms of capital involvement.

However, the use of freezing during at least some stages of cheese production can be of value. Watson and Leighton (1927) showed that the effects of freezing on cheese vary enormously depending on variety, age, moisture and salt content. In general, the texture becomes more crumbly and 'mealy' after thawing, with low moisture content cheeses affected less than those with higher moisture. Thus, cottage cheese, Neufchatel and cream cheese are usually seriously damaged and exhibit whey leakage after thawing, although freezing does not significantly affect cheese flavour.

There has not been a great deal of work reported on the effect of freezing on microbial content of cheese, but, as the activity of micro-organisms is almost entirely stopped by sub-zero temperatures, one use of freezing is to delay the maturation of ripe cheese required for processing. Indeed it is the application of freezing to the storage of all types of cheese and fresh curds for processing that is probably one of the most useful. Burkhalter (1973) showed that freezing of Emmentaler and Gruyère cheese enabled the cheese to be stored for up to 18 months before it was made into acceptable processed cheese, and the only problem was chemical oxidation of fats in the rind resulting in the development of 'tallow' off-flavours. However, the frozen cheeses were not suitable for direct consumption. Filtchakova and Merkulova (1979) reported on a method for the freezing, storing and thawing of cottage cheese that involved the bulk freezing of non-fat curd at below $-30°C$, and storage at a temperature not exceeding $-18°C$. The usual structural damage to the curd was avoided by thawing the frozen curd during the incorporation of the cream. Some cheeses should be frozen before ripening (e.g. Resmini et al. (1973) showed that Provolone curd gave better results when frozen fresh rather than after ripening and matting, as this led to increased proteolysis), but other cheeses, e.g. Gorgonzola (Ottogalli et al. (1973)), can be frozen after ripening with little or no influence on flavour or consistency. Freezing before ripening, however, leads to a breakdown of the casein, a change in the microbial balance, and an acceleration of ripening that gave a bitter flavour and 'mealy' texture. Clearly, almost contradictory results can be obtained by freezing different varieties of cheese.

In Manchego cheese curd, frozen at $-40°C$ and stored for 6 to 12 months at $-10°C$ or $-20°C$, Jimenez Pérez (1978) showed that bacterial counts decreased. However, after thawing and ripening for up to 100 days, the counts rose again to reach normal values, confirming that freezing could, with this cheese, be used as a means of delaying maturation. The relatively small amount of work reported on cheddar cheese (Shannon, 1975) shows that, while the highest numbers of coliforms and lowest total colony counts were observed at −29°C, this situation was reversed at 21°C. Probably the reduction in acid production at lower temperatures favours the survival of coliforms, and hence freezing this type of cheese may have some disadvantages. There was, however, little difference in the texture, body and flavour of the cheeses, except that 'mealiness' was observed in some of the frozen cheese. This disintegration of cheese texture on thawing may be reduced if the freezing process is rapid enough to produce ice crystals of <50 μm diameter (Lück, 1977), but structural changes may still occur if the freezing temperature is too low.

Freezing may also affect the flora of surface-ripened cheeses. Didienne *et al.* (1978) reported on the freezing of some Saint-Nectaire cheese at −30°C. During their investigation, they noted that the counts of lactic streptococci, *Leuconostoc*, lactobacilli and micrococci decreased when the fresh cheese was frozen, but that after thawing, the succession of micro-organisms during ripening was normal. Freezing after three weeks of ripening impaired the development of surface mould, but did not appear to affect the organoleptic quality, and the duration of the freezing similarly had no influence on the final flavour of the cheese.

ICE CREAM

Ice cream, of all frozen dairy products, is the one made and sold in the greatest quantities world-wide. Sales vary enormously from country to country, with the USA and Australia heading the list with over 20 litres per head per annum, to about 6 litres in the UK and 4·5 litres in France.

Ice cream is made by freezing, in a special type of scraped-surface heat exchanger, a mix containing, for example, 10 per cent fat, which may be either butterfat or other suitable edible fat (although some countries, e.g. France, only allow the use of butterfat), 14 per cent sugars, 10 to 11 per cent non-fat milk solids, emulsifier, stabiliser, colour and flavourings. During the freezing, air is incorporated so that the final product contains between 25 and 50 per cent air, although this will depend very much on the mix and the type of freezer being used. After being frozen, the ice cream may be sold either directly from the freezer (soft serve) or, after packaging in bulk (e.g. 1 to 4 litres) or in small unit packages, it may be hardened in a deep freeze or blast tunnel, and sold as the hardened frozen product. Full details of ice cream mixes, their processing and final freezing may be obtained from Arbuckle (1972) or Hyde and Rothwell (1973). Briefly, however, the various ingredients are blended together in the calculated and weighed (or measured) quantities to make a liquid mix. This mix is then subjected to a heat treatment process which, in most countries, is specified by legal requirements so that an adequate reduction in bacterial numbers is obtained, together with the destruction of pathogenic organisms. This heat-treated mix is then homogenised to reduce the size of the fat globules—in order to prevent the fat from churning during the freezing process—and cooled to 4°C or lower for storage until it is frozen and hardened.

The microbiological condition of a product manufactured in this relatively complex manner will depend on the interaction of the following factors:

1. The microbiological standard of the ingredients, and packaging materials.
2. The conditions (e.g. heat treatment) under which the mix has been processed to make ice cream.
3. The hygienic control and cleaning of the equipment used.

The standard of competence of the operators is also critical, but the greater use of automatic control for processes and equipment cleaning has, however, removed some of the problems caused by incorrect operation.

Once the product has been frozen, the microbiological standard is, as it were, locked in, and only very slow changes take place provided that it does remain at a storage temperature of no more than −20°C. At this temperature, there will, in addition, be the minimum deterioration in the physical qualities (such as texture) of the ice cream.

The Raw Materials

Ice cream can be made from a wide variety of ingredients, and, although it is a heat-treated product, it is still necessary to use raw materials of the highest quality, and to store them under conditions which will not allow the proliferation of any micro-organisms.

Milk, cream, skim-milk and skim-milk concentrate, which are used as the ingredients in a very large number of factories, should have been heat-treated, but they must be kept under refrigeration, and be used promptly, to ensure a satisfactory quality. The main organisms present in these dairy materials will be spore-forming bacilli, although there may be some psychrotrophic organisms surviving—if the initial population was high—together with some micrococci and other thermoduric bacteria. Normally, none of these groups constitute a health hazard. Skim-milk and whole milk powders may also contain some bacilli and, it is advisable to ensure, by careful testing of the incoming materials, that their numbers are low; dry, cool storage conditions are essential for these powders. Similarly, with the increasing use of whey powders, including whey protein concentrates, good storage conditions and the use of strict stock rotation are essential.

Butter and butteroil (anhydrous milk fat) are made from heat-treated cream under carefully controlled conditions, and should be of a very high microbiological standard. Spoilage is, therefore, usually chemical rather than microbiological, unless gross contamination by yeasts and moulds has occurred because of careless handling. Similarly, vegetable fats should be of a high microbiological standard, and should be kept, like butteroil, under dry, refrigerated conditions to maintain this high standard. On a large scale, oils and fats are received and stored in a liquid condition at a temperature of around 50°C, and such oils may deteriorate quickly at this temperature; again it is necessary to use them rapidly and in strict rotation. Granulated sugar, glucose syrup solids and dextrose should be almost free of contaminating organisms, but may contain a few yeasts. Similarly, sugar syrups, whether sucrose alone or mixtures of sucrose and glucose syrups, should be virtually sterile, although it should be noted that osmophilic yeasts may be able to grow. Emulsifying and stabilising agents can also prove a hazard unless purchased from a reputable supplier, and kept under good storage conditions.

Nowadays many other ingredients are added to ice cream, or are used as coatings, and as almost all of these are added after heat-treatment of the mix, they constitute a potential hazard. The list includes fruits (canned, fresh or frozen—usually in concentrated sugar syrups), nuts, chocolate, pieces of toffee and biscuit, colours and flavours. Canned fruits should be perfectly satisfactory, but the fresh or frozen equivalent may contain appreciable numbers of yeasts and moulds. Nuts, and in particular desiccated coconut, may contain moulds, and possible mycotoxins, and coconut may also contain salmonellae; careful quality control is necessary to ensure that these products are satisfactory for use. Flavours and colours may be contaminated by careless handling, but, if obtained from a good supply house and stored properly, should not cause problems of microbial origin.

Processing

In all but a very few countries, legal requirements insist on a satisfactory heat-treatment of the liquid mix, including rapid cooling and storage at temperatures below about 7°C until the mix is frozen; some typical heat-treatment requirements are given in Table I.

The bacterial quality of the ice cream mix will depend on the proper operation, maintenance and control of the processing equipment, and although holder processing appears to have a greater safety margin

TABLE I
Some heat-treatment requirements and bacteriological standards for ice cream

Country	Minimum heat treatment requirements	Bacteriological standards
Australia	68°C for 30 min.	Total count max. 50 000 g^{-1}. Coliforms absent in 0·1 g. No pathogenic organisms.
France	Reconstituted or fresh milk must be boiled for 1 min, or heated for 2 to 3 min at 80°C to 85°C, whole mix 60°C to 63°C for 30 min.	Total count max. 300 000 g^{-1}. No pathogenic organisms.
Switzerland	None, but the usual practice is 75°C for 15 to 40 s.	Total count 25 000 g^{-1}. Coliforms absent from 0·1 cm^3. No pathogenic organisms.
UK	(a) 65·5°C for 30 min, (b) 71·1°C for 10 min, (c) 79·4°C for 15 s, (d) 148·8°C for 2 s. All followed by cooling to below 7·2°C within 1·5 h of heating.	None: a methylene blue reduction test is used by public health authorities as a guide to hygiene quality.
USA	Individual states have different requirements, e.g. 68·3°C for 30 min, 79·4°C for 25 s.	Total count max. 50 000 to 100 000 g^{-1}.

than short-time processing, it is essential that the time and temperature conditions should be carefully adhered to. Over-heat treatment should be avoided as this may lead to undesirable flavour changes. The whole process, including cooling and the storage of the mix (at temperatures between 3°C and 7°C), should be carefully controlled and monitored by management. In addition, it is important to note that ice cream mix has been shown (Singh and Ranganathan, 1978) to exert a greater protective influence on organisms than liquid milk. These authors showed that the D-values at 50°C and 55°C of *Escherichia coli* were approximately double those found when using cows' milk.

The freezing process, during which the ice cream mix is subject to considerable agitation as well as a reduction in temperature in a specially designed, scraped-surface heat exchanger, also causes changes to occur in the microbial content of the product. During the freezing operation, air is whipped into the mix, and the resultant foam is stabilised by freezing a substantial part of the water in the mix into very small ice crystals.

Some types of freezer actually bring about a reduction in the microbial content of the ice cream. Thus, Alexander and Rothwell (1970) showed that, in a small vertical freezer in which the freezing process took about 10 to 15 min, a remarkable drop occurred in the numbers of viable *Escherichia coli* and *Bacillus cereus* organisms which had been added to the mix prior to freezing; the very rapid freezing which occurs in a continuous freezer, however, caused a much smaller reduction. The destruction of the organisms appeared to begin when ice crystal formation began, and mechanical agitation was also necessary; bacterial numbers in mixes frozen under quiescent conditions showed little or no change. Similarly, the rate of microbial destruction did not appear to alter with increased total solids. However, the freezing process must never be relied upon to reduce the bacterial content of the final product.

The ice cream may be sold direct from the freezer, as a 'soft-serve' product, or it may be further reduced in temperature, in a hardening room or blast-freezer tunnel, to produce 'hard' ice cream. The temperature of this latter product will be reduced to below−20°C, and the ice cream will be stored at this temperature until it is sold. Even if the period between freezing and final sale is several months, there will be little change, if any, in the microbial content of the ice cream. An extensive literature ranging from Abd-El-Bakey and Zahra (1978) to Wallace and Crouch (1933) shows that both *Mycobacterium* and

Salmonella, as well as many other less harmful types, can survive at
−20°C for very long periods. They do not multiply, provided that the
temperature is low enough for the ice cream to remain hard, so that in
effect, the microbial quality of ice cream is 'locked-in' by the hardening
process. It has to be stressed, therefore, that the quality of the product
will not be altered by the final hardening, but depends on the standard
of hygiene of the equipment and the quality of the processing. Low
bacterial content of ice cream coming from the freezer is essential, and
demands the utmost care and attention to the details already
mentioned.

There have been cases of food poisoning caused by ice cream. In the
UK, however, there have been none reported since 1955—a state of
affairs brought about by the diligence of the Public Health Authorities,
the care of the processors, and an increasing awareness throughout the
industry that hygiene must be a prime concern. The last major
outbreak was, in fact, in 1947 (Evans, 1947), when about 210 cases of
typhoid fever, including four deaths, were reported from Aberystwyth.
The incident was caused by ice cream being contaminated by the
manufacturer who was a urinary excretor of *Salmonella typhi*; the
manufacturer had been declared free of the infection in 1938. Other
cases reported from the UK (Hobbs and Gilbert, 1978) include
paratyphoid, *Shigella* dysentry caused by an ice cream being accidently
touched by a monkey, eleven outbreaks involving various salmonellae,
two involving staphylococci and six of unknown cause. Even when the
greatest care has been taken to produce an ice cream of the highest
quality, it is still liable to contamination at the final point of sale. One
particular hazard arises when 'soft-serve' ice cream is being dispensed,
and another is caused by selling, from bulk containers, single portions
as cones or wafers.

OTHER FROZEN DESSERTS

There is a growing market worldwide for frozen desserts based, to a
greater or lesser extent, on milk ingredients. Some of these products
are quite complex, while others are relatively simple, and the range
available includes ice lollies, sherbet and water ice, mousse, and
complex desserts incorporating, for example, ice cream, whipped
cream, sponge cake, chocolate and fruit and nuts.

Water ices, ice lollies and sherbets, most of which are low pH items, should be of a high microbial standard, provided that they are not grossly contaminated by careless handling. Mousse, which is generally understood to be a high fat (either vegetable or milk fat), highly stabilised and highly whipped product, is normally produced in a similar manner to ice cream. One method of manufacture incorporates the highly concentrated gel section of the mix into a separately produced mix very similar to ice cream, and both of these fractions will have been subjected to heat treatments similar to those given to ice cream mixes. If they are then cooled rapidly, combined together and frozen without delay, the resulting product should have a similar microbial standard to ice cream. The desserts made by incorporating ice cream, cream, sponge cake and other additions will be handled much more than the products mentioned so far in this section, and such confections are, therefore, much more susceptible to contamination. Despite this risk, there appears to be no reference in the literature to any outbreaks of food poisoning being caused by these types of product.

REFERENCES

Abd-El Bakey, M. A. and Zahra, M. (1978). Research Bulletin, Faculty of Agriculture, Ain Shams University, Egypt.
Alexander, J. and Rothwell, J. (1970). *Journal of Food Technology,* **5,** 387.
Anon. (1979). *Milk Industry,* **81**(5), 10.
Arbuckle, W. S. (1972). *Ice Cream,* AVI, Westport, Connecticut.
Babcock, C. J. Roerig, R. N., Stabile, J. N., Dunlap, W. A. and Randall, R. (1946). *Journal of Dairy Science,* **29**(10), 699.
Babcock, C. J., Roerig, R. N., Stabile, J. N., Dunlap, W. A. and Randall, R. (1947). *Journal of Dairy Science,* **30**(1), 49.
Babcock, C. J., Strobel, D. R., Yager, R. H. and Windham, E. S. (1951). *Journal of Dairy Science,* **34**(6), 488.
Bashandy, E. Y. and Heider, L. E. (1979). *Zentralblatt für Veterinär-Medizin,* **B26**(1), 1.
Bell, R. W. J. (1948). *Journal of Dairy Science,* **31**(11), 951.
Burkhalter, G. (1973). *Schweizerische Milchzeitung,* **99**(6:7), 33:41.
Chase, D. (1981). *Milk Industry,* **83**(6), 18.
Coghill, D. M. and Juffs, H. S. (1978). *Dairy Products,* **6**(1), 5.
Davis, J. G. (1981). In *Dairy Microbiology, 2,* (Ed. R. K. Robinson), p. 31, Applied Science Publishers, London.
Didienne, R., Millet, L. and Verlaguet, C. (1978), XXth International Dairy Congress Vol. E 1011.
Dixon, B. D. and Rochford, J. N. (1973). *Australian Journal of Dairy Technology,* **28**(2), 67.
Evans, D. I. (1947). *The Medical Officer,* 25th January, 39.

Filtchakova, N. N. and Merkulova, N. N. (1979). *Bulletin de l'Institut International du Froid,* **59**(4), 1188.

Gebre-Egziabher, A. (1982). *Journal of Food Protection,* **45**(2), 125.

Gilliland, S. E. (1977). *Journal of Dairy Science,* **60**(5), 805.

Giroux, R. N., Martin, C. and Samson, R. (1971). *Canadian Institute of Food Technology Journal,* **4**(2), 55.

Hobbs, B. C. and Gilbert, R. J. (1978). *Food Poisoning and Food Hygiene,* 4th Edition, Edward Arnold, London.

Hyde, K. A. and Rothwell, J. (1973). *Ice Cream,* Churchill Livingstone, Edinburgh.

IDF (1974*a*). Document 83. Bacteriological quality of cooled bulk milk. International Dairy Federation, Brussels.

IDF (1974*b*). Document 82. Lypolysis in cooled bulk milk. International Dairy Federation, Brussels.

IDF (1980), Document 120. Factors affecting the bacteriological quality of raw milk. International Dairy Federation, Brussels.

IDF (1981), Document 130. Factors affecting the keeping quality of heat treated milk. International Dairy Federation, Brussels.

Jimenez Pérez, S. (1978). *Mimentaria,* **96, 97, 98;** 23, 21, 35.

Kastornykh, M. S., Khomutov, B. I. and Gumennaya, I. P. (1976). *Nauchnye Trudy Moskovskogo Instituta Narodnogo Zhozyaĭastva,* **5,** 71.

Kratochvil, L. and Vedlich, M. (1961). *Zpravy Vgz Kumneho Ustavu Mlekarenskeho, Praha,* **9**(1), 26.

Löndahl, G. and Åström, S. (1979). *Frozen Foods,* **32**(3), 39.

Lück, H. (1977). *South African Journal of Dairy Technology,* **9**(4), 127.

McDowall, F. H. (1953). *Buttermaker's Manual,* 2 vols, NZ University Press.

McKinney, J. D., Cavanagh, G. C., Bell, J. T., Hoversland, A. S., Nelson, D. M., Pearson, J. and Selkirk, R. J. (1973). *Journal of the American Oil Chemists Society,* **50**(3), 79.

Manu, V., Tofan, M. and Fromunda, M. (1972). *Industria Alimentara,* **23**(12), 679.

Muir, D. D., Phillips, J. D. and Dalgleish, D. G. (1979). *Journal of the Society of Dairy Technology,* **32**(1), 19.

Murphy, M. F. (1981) In *Dairy Microbiology 2* (Ed. R. K. Robinson), p. 91, Applied Science Publishers, London.

Murray, J. G. and Coey, W. E. (1959). *Journal of Applied Bacteriology,* **22**(1), 125.

Osborne, R. J. W. (1977). *Journal of the Society of Dairy Technology,* **30,** 40.

Ottogalli, G., Resmini, P., Rondinini, G. and Volonterio, G. (1973), *Scienza e Tecnologia degli Alimenti,* **3**(1), 43.

Randall, R. (1949). *Refrigeration Engineer,* **57**(9), 883.

Resmini, P., Volonterio, G., Piergiovann, L. and Bernardi, G. de (1973). *Latte,* **47**(4), 252.

Richards, E. (1982). *Journal of the Society of Dairy Technology,* **35**(4), 149.

Samuelsson, E.-G., Thomé, K. E., Borgström, G. and Hjälmdahl, M. (1957). *Dairy Science Abstracts,* **19**(ii)8, 75 (Review Article No. 65).

Schulz, M. E. (1964). *Milchwissenschaft,* **19**(1), 9.

Shannon, C. W. (1975). *Dissertation Abstracts International,* **B35**(12), 5935.

Sims, J. E., Kelley, D. C. and Foltz, V. D. (1969). *Journal of Milk and Food Technology,* **32**(12), 485.

Singh, R. S. and Ranganathan, B. (1978). XXth International Dairy Congress, Vol. E, 855.

Stadhouders, J., Jansen, L. A. and Hup, G. (1969). *Netherlands Milk and Dairy Journal,* **23,** 182.

Storper, M., Ziu, G. and Saran, A. (1982). *Refuah Veterinarith,* **29**(1/2), 6.

Tamime, A. Y. (1981). In *Dairy Micriobology, 2* (Ed. R. K. Robinson), p. 135, Applied Science Publishers, London.

Wallace, G. I. and Crouch, R. (1933). *Journal of Dairy Science,* **16,** 315.

Watson, P. D. and Leighton, A. J. (1927). *Journal of Dairy Science,* **10,** 331.

Wearmouth, W. G. (1957). *Dairy Engineering,* **74,** 193.

Yano, N. and Morich, T. (1971). *Journal of the Food Hygiene Society of Japan,* **12**(5), 408.

Chapter 7

Freezing for the Catering Industry

R. K. Robinson

Department of Food Science, University of Reading, UK

INTRODUCTION

The second half of the twentieth century has brought with it some dramatic changes in the patterns of food consumption in the industrialised countries. One of the most notable of these changes has been the acceptance by all age groups of routines that involve the eating of meals prepared outside the home. Children have lunches at school; factory and office workers are provided with canteen facilities; and many others make regular use of cafés and restaurants for business or social purposes.

Initially, this trend was handled by an extension of the 'domestic' approach to meal preparation, and most lunch-time offerings would have consisted of freshly cooked meat/fish and vegetables served with the minimum of delay after cooking. However, this traditional approach could not meet the realities of the new situation, and in particular:

(i) the demand for large numbers of meals per day in institutions, such as hospitals or canteens;
(ii) the need to offer more extensive and varied menus;
(iii) the necessity of handling the day-to-day fluctuations in demand observed in restaurants and cafés; and
(iv) the almost universal difficulty of handling massive peaks in demand at certain times of the day.

The inevitable solution to this advancing chaos was to invoke such

233

TABLE I

Outbreaks of food poisoning associated with the preparation of meals on a large scale; figures for the years 1976–78

Source	Causal organism		
	Salmonella	Staphylococcus aureus	Clostridium perfringens
Hospital	127	4	47
Restaurants	83	3	33
'Celebrations'	70	2	9
School canteens	13	6	27
Institutions	14	6	42

After Hepner (1980).

dubious expedients as storing large quantities of prepared food for subsequent reheating, or putting out large numbers of cold dishes that reached room temperature long before consumption, and the outcome was as predictable as it was serious. Thus the figures shown in Table I are merely for *recorded* incidents of food poisoning, and hence the *actual* number of individuals who have suffered some degree of debility from consuming ill-prepared food must be many times greater. Indeed a recent report suggested that some 23 million working days were lost during 1980 as a result of food poisoning, and although this figure is a fairly crude estimate, it does give an indication of the magnitude of the problem.

Although the catering industry was conscious of these unfavourable statistics, it was the sheer mechanical inefficiency of the traditional approach to the provision of large numbers of meals that prompted caterers to evaluate their operational deficiencies. The solution seemed to lie in the direction of employing more factory-processed foods and, for reasons of quality and shelf-stability, frozen products featured strongly on the revitalised menus.

In general, this initial introduction of frozen foods involved the use of:

(i) items, such as out-of-season vegetables, or fish and poultry, that could be purchased in bulk, and then used as required to extend a basic menu; and

(ii) ready-prepared components of a meal, e.g. pizzas or meat pies, whose introduction involved the kitchen in minimal preparation before serving.

This trend solved both the problem of restricted menus and, at least to some extent, the difficulty of accommodating unexpected arrivals, but the handling of peak demands still provided a major headache for the industry. It was perhaps logical, therefore, that the next stage in the evolution should involve the purchase of entire frozen meals, and the result was a totally new approach to large-scale catering—the cook–freeze system.

In this system, complete meals are prepared and portioned out, and the entire serving is then frozen until required. This concept has a number of significant implications, and in particular, it should be noted that:

(i) the company serving the food and the company preparing the meals can be totally independent and, even if associated, the primary kitchen can be spatially separate from the main serving area; and

(ii) there is no longer any temporal link between preparation and consumption, and hence problems of peak demand can be handled with less pressure on catering staff.

The contrast between the conventional approach to catering and a

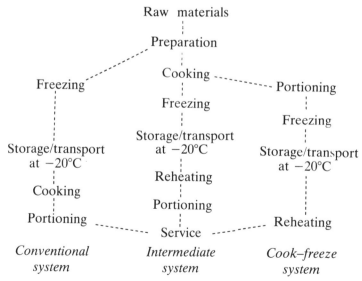

Fig. 1. The principle routes by which frozen foods are employed for the production of meals in restaurants or canteens. After Glew (1977).

cook–freeze operation has been discussed in detail by Glew (1973), but, in the present context, it is relevant that the cook–freeze system is only one of the ways in which frozen foods can be employed in the catering industry. Thus, as indicated earlier, frozen products are now incorporated into service operations in a number of ways, but the main pathways can be summarised as in Fig. 1.

Each of these basic avenues has its supporters, and the selection of any one system will depend largely on the scale of operation involved. Similarly the microbiological hazards are a reflection of the method of handling, and although the divisions are in no sense intended to be rigid, it is convenient to consider them on an individual basis.

CONVENTIONAL SYSTEM

Many cafeterias still rely on a traditional approach to the preparation of meals, and the only real acknowledgement of recent developments has been the switch to more convenient raw materials. Dried products like peas were the original pacemakers in this direction, but the superior organoleptic quality of frozen foods gave them the decisive edge. In addition, the variety of available foods offered the restauranteur an opportunity to provide a comparatively exotic menu, for, as can be seen from Table II, the range of basic ingredients to be found in a deep-freeze is now considerable.

Although some of these items, such as peas or beans, will have been blanched prior to freezing, the essential feature of these foods is that they are raw, and this factor has a number of implications in that:

   (i) the food will, after thawing, be subject to a cooking process which should eliminate many of the microbial contaminants, although some toxins are able to survive both heating and freezing;

  (ii) in many cases, the natural, saprophytic microflora on the surface of the food will be intact, so that during thawing, the growth of any pathogens deposited on the food may be restricted;

 (iii) although freezing will damage the living tissues of a foodstuff, e.g. raw meat, it may still retain some of its inherent resistance to microbial growth; and

 (iv) freezing at −18°C or below kills many of the organisms that were initially present.

TABLE II
A selection of uncooked meal items that are now available in freezer
packs specially prepared for the catering trade

Group	Example
Fish	Salmon (headless, gutted)
	Rainbow trout
	Dover/lemon soles
	Fillets of plaice, cod and haddock
Meat	Steaks
	Pork and lamb chops
	Minced/diced beef
	Chicken (whole and portions)
	Turkey
	Duckling
	Guinea fowl
Vegetables	Peas
	Beans
	Sprouts
	Carrots
	Cabbage
	Chipped potatoes
Fruit	Blackcurrants
	Raspberries

Courtesy of Brake Bros. (Frozen Foods) Ltd.

Thus in general terms, raw, frozen produce should provide a
satisfactory starting point for the preparation of meals of sound
microbiological quality, and yet incidents of food poisoning still arise.
The reasons for these outbreaks can be summarised as:

(i) the presence of pathogenic organisms in the foodstuff, e.g.
already associated with an animal prior to slaughter, or derived
from the factory premises during the preparative, freezing or
packing operations;
(ii) contaminated surfaces or equipment in the kitchen of the
catering establishment; and
(iii) conditions favouring microbial growth and/or metabolism in the
period between defrosting and serving to a susceptible human.

The relative importance of these three areas can be judged from
Table III, and while neglect of one or more of these factors can cause

TABLE III
Some of the factors that contributed to the outbreaks of food poisoning
reported in England and Wales for the years 1970–79

Factor involved	Number of incidents
Contaminated processed food	199
Inadequate thawing	64
Undercooking	161
Inadequate cooling	333
Storage above 10°C	413
Inadequate serving temperatures	60
Infection from other foods	62
Infection from food handlers	54

After Harrigan (1983).

problems with any form of food preparation, their specific relevance to
frozen foods can be visualised from the following example reported by
Hobbs and Gilbert (1978).

Frozen Minced Beef

Although whole cuts of meat should pose few problems from the
microbiological standpoint, minced meat (see Chapter 4) is invariably
contaminated during the process of maceration. A range of genera may
be represented in the final product, but the ubiquitous distribution of
Clostridium perfringens makes it a likely candidate for inclusion.

It was perhaps no surprise, therefore, when an outbreak of food
poisoning was traced to some frozen minced beef. The meat in
question was delivered in 2·7 kg frozen blocks, and six such blocks
were heated in stock in a large pan. After cooling at ambient
temperature over a period of some 5 h, the mince was transferred to a
refrigerator for holding overnight. Next morning the meat was
reheated and then distributed as individual portions in food containers.
The containers were subsequently stacked in insulated, electrically
heated cabinets of a van used for distribution. Although the cabinet
was designed to hold the food above 60°C, in practice many of the
meals did not even reach 50°C, and many could have been
considerably cooler. The fact that the delivery took 3–4 h completes
the saga, and the various faults that led to the outbreak of food

poisoning become, in retrospect, all too clear:

(i) the rate of heat transfer through a block of frozen meat is inevitably slow, and hence it is likely that the blocks of frozen mince were not heated throughout. As a consequence, spores of *Clostridium perfringens* remained intact;

(ii) the slow cooling process would have given these same spores ample time to germinate; and

(iii) the mild reheating procedure would not have been sufficient to kill the vegetative cells released by the developing spores, and the subsequent holding of the meals at 40–50°C would have provided near ideal conditions for the growth of the organism.

The problems that can arise from the slow cooling of animal products, inadequate reheating and/or holding at ambient temperatures are ones that can be associated with any form of catering, but the phenomenon of inadequate heat penetration of frozen products is entirely specific. Small items can, of course, be cooked directly without thawing, e.g. peas or fish fingers, but large blocks of mince, frozen chicken or turkeys, or joints of meat—especially those that have been rolled so that the contaminated surface is in the centre—must be thawed completely prior to cooking. Failure to do so means that portions of the food may still be frozen at the time that the joint or carcass is placed in the oven, and it is unlikely that such regions will ever reach a temperature that is lethal to bacterial spores. In some cases, the temperatures attained will not even destroy vegetative cells, and indeed the almost ritual outbreaks of food poisoning at Christmas, following the consumption of turkey contaminated with *Salmonella* spp., are ample testimony to the ease with which a poorly defrosted carcass can harbour living bacteria throughout a normal cooking process.

Nevertheless, raw frozen produce should be totally acceptable as a basis for the production of meals on a large scale, always bearing in mind that:

(i) the foods must be purchased from a reliable supplier, so that their microbiological quality is, prior to freezing, of a high standard;

(ii) adequate refrigerators or thawing cabinets must be available for handling large items;

(iii) the preparative procedures employed in the kitchen must be

TABLE IV
Some common sources of the food-poisoning organisms indicated (in some
cases, contamination of the frozen food is unavoidable, and hence correct
procedures in the kitchen are essential to avoid placing consumers at risk)

Organism	*Foodstuff*
Salmonella	Poultry, sausages, meat pies
Clostridium perfringens	Minced meat, meat (pies)
Staphylococcus aureus	Cream desserts and cakes, cold meats
Bacillus cereus	Spices and seasonings
Clostridium botulinum (E)	Raw shellfish and seafoods

designed to minimise contamination and the growth of micro-
organisms; and

(iv) the physical lay-out of the kitchen, as well as the available
equipment, must be conducive to hygienic operation (Boltman,
1978).

These points apply with equal force to users of all types of
potentially suspect frozen foods (see Table IV), and hence attention
can now be focussed on the special problems associated with
pre-cooked frozen foods.

INTERMEDIATE SYSTEM

This category is intended to include all those pre-cooked and
ready-prepared foods that only need to be heated and portioned out
prior to consumption. Some familiar examples are shown in Table V,
and it is significant that:

(i) although the microflora will have been reduced by the initial
cooking process, the foods are not intended to be 'sterile';
(ii) the partial loss of the normal surface population of the foodstuff
may encourage proliferation of an undesirable species that is
able, in the absence of competition, to colonise the available
space;
(iii) the foodstuff will have lost most of its natural resistance to
microbial invasion; and
(iv) the final heating will be carried out solely to render the food fit
for consumption, and it will, at that point, be anticipated that
the item is microbiologically 'safe'. In other words, the

TABLE V
Some examples of ready-prepared components of a meal
that are pre-cooked, at least in part, prior to freezing

Group	Example
Seafood/fish	Scampi Lobster in brine Cod/haddock in batter
Meat	Burgers in buns Chicken (whole and portions) Turkey (whole and portions)
Including meat	Meat pies (various) Cornish pasties Pizzas (various)
Vegetables	Oven chips Roast potatoes Potato croquettes Sauté potatoes Baked jacket potatoes

Courtesy of Brake Bros. (Frozen Foods) Ltd.

consumer/user may well assume that merely 'warming' the food is sufficient.

This latter attitude means that the microbial quality of the product should, prior to freezing, be excellent.

In practice, attention to detail in this area becomes part of the general necessity of providing a 'wholesome' food, and this concept was defined by Robson (1978) as the provision of a product that will:

(i) satisfy the reasonable expectations of a consumer, where these expectations are based on previous experience or advertising claims;
(ii) meet the legal requirements of the country in which it is sold;
(iii) be manufactured under hygienic conditions, in that both the building and the equipment can be maintained to a satisfactory standard; and
(iv) contain minimal levels of hazardous components, e.g. micro-organisms, toxins or environmental pollutants.

In the present context, it is points (iii) and (iv) that are of critical importance, but, because equipment design and hygiene are relevant to all aspects of catering, it is the problem of contamination that is of immediate relevance. This introduction of foreign materials can arise at a number of points, e.g. low-grade raw materials, poorly designed processing and filling operations, or faulty handling during storage, but it may be of specific interest that frozen prepared foods are often chemically modified *vis-à-vis* their normal counterparts. Thus Hill (1977) described prepared frozen dishes under four separate headings:

1. Foods that can be cooked, frozen and then reheated without any detectable change in their basic properties, e.g. meat without gravy, cakes and pastry, as well as simple components like boiled rice or spaghetti.
2. Products that can only be rendered satisfactory by modifying the recipe and/or the processing conditions, e.g. thick sauces, egg dishes, as well as milk puddings and cold desserts.
3. Items that are acceptable in the short-term, but deteriorate rapidly during storage, e.g. boiled bacon.
4. Foods that never appear on a supplier's list, namely entire cooked eggs.

Care in the handling of all these categories is obviously essential, but it is pertinent to consider whether any of the modified processes and/or additional ingredients introduced into foods in Group 2 could pose additional microbiological hazards.

Cooked vegetables should cause few problems, for, as the cooking process is designed to inactivate the natural enzymes in the food, the vegetative cells of bacteria should be similarly destroyed (Ayres *et al.*, 1980). However, the short time for which vegetables are immersed in boiling water, e.g. two minutes for cauliflower and three minutes for cabbage will not be sufficient to remove spore-formers. Spore-formers may similarly be present in some of the thick, starch-rich sauces employed in pies and other products, but the influence of these spores on the quality of the finished product will depend, in large measure, on the handling of the food during thawing and reheating; however, their presence could be disadvantageous.

These rather specific sources of contamination are, of course, additional to those associated with any food processing plant, and hence serve to accentuate the anticipated bacterial load. Obviously this total aerobic flora cannot be taken as a direct measure of acceptability of a frozen food (Dack, 1956), but it can prove a valuable indicator of sanitation. In addition, Hobbs (1965) put forward the more general

TABLE VI

Some bacterial counts obtained by an examination of frozen, uncooked chicken and turkey pies; all figures as average number of micro-organisms per gram

Brand	Total count[a]	Coliforms	Staphylococcus (coagulase +)
1. Chicken	13 700 (3 000)	5 (0)	400 (—)
Turkey	30 700	38	10 400
2. Chicken	1 804 000 (10 000)	560 (0)	108 000 (*)
Turkey	1 777 500	630	3 800
3. Chicken	362 500 (4 200)	90 (0)	12 250 (*)
Turkey	84 500	5	7 750

[a] The figures in brackets represent a typical count after the pie has been cooked in accordance with the manufacturers instructions.
(—) No test performed.
(*) Variable numbers of *Staphylococcus* survived.
After Canale-Parola and Ordal (1957).

point that most foods involved in food poisoning were those with large microbial populations overall, and these two conclusions have given total colony counts a respectibility that is not always justified (Hill, 1983). Nevertheless, the advent of automated techniques for performing total colony counts has tended to expand their usage rather than diminish it, and their application to monitor pre-cooked frozen foods is no exception.

Some typical results relating to frozen pies and other main course dishes are shown in Table VI, and it is likely that these figures are fairly typical. However, Goldenberg and Elliott (1973) suggested that meat pies could be produced for retail with total colony counts as low as $200 \, \text{g}^{-1}$, and it does raise the question as to whether even more stringent specifications could not be achieved.

COOK–FREEZE CATERING SYSTEMS

The most convenient of all the systems as far as the caterer is concerned is the cook–freeze system, in that by separating, in time and space, the cooking and serving operations, the historical pressures on staff are largely removed. Thus, the routine activities in a kitchen may be largely reduced to reheating and serving, or, for cold dishes, thawing and serving, but, as indicated earlier, even these procedures must be handled with care. Nevertheless, the fact that the kitchen and the canteen are now effectively unit operations, makes it convenient to consider the problems of hygiene as they affect the key area of operation—the kitchen.

Hygiene in the Kitchen

Although the code of hygiene for any food preparation should be strict, the pressure is less intense if the food is to be thoroughly cooked immediately prior to consumption. Where mild reheating or even thawing are the only treatments involved, then it is imperative that cleanliness in the production unit is beyond criticism. Efficient management and training can, of course, go a long way to achieving this goal, but the environment of the workplace can negate even the best of intentions. In some establishments, the constraints of the existing building make the introduction of hygienic practices extremely taxing, especially if it is decided to accommodate the entire cook–freeze–serve operations within the same kitchen/servery area. If meal production and freezing can be handled in a separate unit, then modernisation of the reheating/dispensing facility should not be a problem, for then the overall lay-out of the production area becomes, in both operational and microbiological terms, the central issue.

An excellent review of the main considerations was presented by Furnivall (1977), but those aspects that mainly influence the microbial load on the meals under preparation can be summarised as:

(i) Storage facilities must be able to cope easily with the reserves required; all stores should be cool and well ventilated; all shelves and bins must be readily accessible, and must be of a construction that is easily cleaned.

(ii) The floors in all areas must be smooth, and impervious to water, detergents or grease, e.g. quarry tiles; glazed tiles make an equally efficient covering for walls. It is essential that the fixing is carried out by a reputable company, because poorly laid tiles can soon become a haven for innumerable micro-organisms.

(iii) Whether the meals are frozen on-site or prepared elsewhere, the siting of the refrigeration and deep-freeze units should be considered at an early stage. Convenience of operation is, of course, a major consideration, but projected requirements in terms of capacity are an equally important aspect. Thus, the desirability of separating cold desserts from other foods is an obvious point, but equally relevant is the fact that an overstocked unit in constant use will be hard-pressed to maintain cold conditions as the ambient temperature rises.

The location of the associated compressors also deserves more attention than it usually receives, because, aside from any mechanical considerations, most units act as unintentional traps for dust and grease, and hence provide a constant reservoir for moulds.

(iv) The dimensions of any food preparation area will vary considerably with the organisation, but where different foods are handled, the oft-quoted rules relating to the need to isolate raw foods from cooked foods, or different types of raw food from each other, e.g. meat and vegetables, still apply.

(v) It is probably only natural that discussions of food hygiene centre on the food and its preparation, but the linked activities of waste disposal and dishwashing can, if neglected, provide a major source of infection. Thus, a plate coated in gravy or custard can be as effective as a petri dish of agar in supporting microbial growth, and, if left in an exposed position, the risk of cross-contamination is self-evident.

It is important, therefore, that the dishwashing area should be located with direct access to the dining area, and should be well away from the sections associated with food preparation. An increased use of mechanisation is also desirable in this area, not only through the benefits of saving on labour, but also because the higher washing temperatures employed, and the more rapid throughputs, will minimise the risks of contamination. A typical, semi-automated area is shown in Fig. 2, and under such conditions the spread of microbial contaminants should be minimal. In large canteens, the rapid return of crockery and cutlery to the washing area may be facilitated by mechanical aids, such as a helical conveyor belt, but, whatever the chosen system, the rapid cleansing of the dirty dishes is essential.

The rapid discard of plate waste via a disposal unit is clearly efficient from the microbiological point of view (see Fig. 3), but the same attitude rarely extends to the handling of swill bins.

Fig. 2. A modern 'washing-up' unit designed for both efficiency of operation and a high level of hygiene. (Courtesy of Meiko.)

Fig. 3. In this unit, plate waste is transferred to a water chute that removes the debris from the kitchen area. (Courtesy of Meiko.)

All too often these latter receptacles are simply placed outside the kitchen door on floors that are not amenable to hosing-down or sweeping, while the bins themselves are often 'less than sanitary'. Obviously the cleaning of this area is not one of the more attractive tasks on the premises, but it is, nonetheless, one that should be performed routinely and with diligence.

(vi) The general 'atmosphere' within a kitchen is a facet of the operation that is not strictly within the scope of this chapter, except in so far as it influences attention to detail. Thus, nobody working in conditions of poor lighting, poor ventilation and excessive heat can be expected to pay much attention to the elements of food hygiene, and even the most competent of staff may succumb to the pressures of an unhelpful environment.

If a canteen/restaurant kitchen is undergoing a comprehensive reconstruction, as might be the case if an on-site cook–freeze system is being introduced, then improving the lay-out of the premises and/or the general working environment is clearly feasible. More restricted developments may, however, have to reach a compromise between the conflicting demands of reality and 'the ideal', but one area in which cost-saving is truly a false economy concerns the actual equipment in the kitchen.

Hygienic operation of kitchen equipment

It is stating the obvious to highlight the need for all equipment and utensils to be of a high standard of hygiene, but where meals are being portioned for freezing or merely being reheated for consumption, microbiological considerations achieve a special significance.

The main problem encountered, in this context, is that kitchen equipment is often designed solely with a specific cooking operation in mind, or even with the aesthetic character of the apparatus as the priority, and the elimination of 'traps' for dust or dirt is a somewhat secondary consideration. The need to 'design for hygiene' is, however, gradually being recognised, and this trend will become even more important if, for reasons of economy, the quality of the materials used in construction declines. Thus, a change to a lower grade of steel, for example, may allow excessive flexing, and the presence of hollows in which a residual film of food can escape detection. The use of pop-rivets can similarly reduce costs, but often at the expense of gaping seams in which bacteria can flourish. If the finish becomes pitted and uneven, cleaning again becomes difficult, but it should not be forgotten that it is the overall design that establishes, to a large extent, the attitude of the kitchen staff to cleanliness. In other words, if a piece of equipment is simple to dismantle and clean, it will be sanitised regularly and thoroughly, whereas difficult operations will tend to be 'put-off' until sheer necessity determines otherwise. The accessibility of the food contact surfaces is a vital point, but equally important is the fact that they should be engineered with hygiene in mind.

The essential of smooth surfaces has already been mentioned, but if the lay-out provides dead-spaces, crevices or even acute corners, then there is every chance that food particles will become lodged at the offending points and support bacterial growth. The ability of a piece of equipment to drain efficiently is a further aspect that must not be neglected, for otherwise a residual 'pool' will, overnight, become a serious source of contamination.

The design of the plant is clearly critical for achieving hygienic operation, but the actual procedure of cleaning can have an equal bearing on the restraint of bacterial loads. The sequence of operations for cleaning and the choice of active agents will vary according to the type and scale of the operation, but, in general, a programme might involve:

(i) rinsing the equipment with warm water in order to remove any food residues;

(ii) dismantling those sections of the apparatus that are in contact with the food;

(iii) scrubbing the relevant pieces with hot detergent solution; and

(iv) after re-assembling, rinsing the equipment with clean water.

A sanitising procedure involving hot water or a proprietary chemical agent is an obligatory follow-up to the general cleaning operation, and whether the sanitising solution is circulated through the equipment or the entire item is immersed, an adequate contact time must be observed. If a chemical sterilant is employed, then a final rinse in clean water will be required to avoid the danger of chemical taints, but even if hot water alone has been used, the unit must still be drained thoroughly and allowed to dry.

It is intended, of course, that the above sequence should become a defined routine, but what is equally important is the need to ensure that operatives appreciate, at least in general terms, the intention of each stage. This stricture applies with equal force to many aspects of

TABLE VII
Some possible causes of bacterial growth/metabolism on foodstuffs, and an
indication of the requirements for prevention

Activity	*Risk*	*Remedy*
Food Storage	Microbial activity	(i) Check temperature of deep-freezes at regular intervals. (ii) Do not use storage area to freeze foods from ambient.
Initial preparation	Dirty hands	(i) Wash thoroughly with soap and hot water. (ii) Wear disposable gloves and use for one operation only. (iii) Do not touch surfaces that may be contaminated.
	Dirty surfaces and utensils	(i) Clean thoroughly with hot detergent solution, and with a proprietary sanitising agent. (ii) Wash equipment before switching to a different use.
Cooking	Inadequate heating	(i) Thaw products completely. (ii) Allow sufficient time for the food to attain an adequate temperature throughout.
Cooling	Microbial activity	(i) Cool as rapidly as possible. (ii) Once temperature of food reaches ambient, refrigerate promptly. (iii) Check temperature of refrigerators at regular intervals.
Serving	Contamination	(i) Keep dirty plates, etc., separate from outgoing meals. (ii) Maintain high standard of hygiene among operatives. (iii) Avoid contaminination between foods.
	Microbial activity	(i) Serve food hot (65°C or above) or cold (4–5°C).

kitchen/servery, and Table VII indicates some typical faults that arise because personnel have no knowledge of, or do not think about, the microbiological implications of their actions.

The extent to which the cook–freeze system of catering will become popular remains to be seen, but there is no doubt that frozen foods as a whole have had a dramatic impact on the industry. The benefits to the caterer and the consumer are self-evident, but the technology of thawing and heating frozen items is still, judging by the evidence in Table III, fairly primitive. The more widespread use of microwave units may help to take some of the 'guesswork' out of the defrosting stage, but for the present, it is essential that kitchen staff are made more aware of the microbiological implications of their actions. If this improvement can be achieved, then frozen foods should become amongst the 'safest' for large-scale catering.

REFERENCES

Ayres, J. C., Mundt, J. O. and Sandine, W. E. (1980). *Microbiology of Foods,* W. H. Freeman & Company, San Francisco.

Boltman, B. (1978). *Cook–Freeze Catering Systems,* Applied Science Publishers, London.

Canale-Parola, E. and Ordal, Z. J. (1957). *Food Technology,* **11,** 578.

Dack, G. M. (1956). *Food Poisoning,* University of Chicago Press, Chicago, Illinois.

Furnivall, M. E. (1977). In *Catering Equipment and Systems Design* (Ed. G. Glew), Applied Science Publishers, London.

Glew, G. (1973). *Cook–Freeze Catering Systems,* Applied Science Publishers, London.

Glew, G. (Ed.) (1977). *Cook–Freeze Catering: an introduction to its technology,* Faber and Faber, London.

Goldenberg, H. and Elliott, D. W. (1973). In *The Microbiological Safety of Food* (Eds B. C. Hobbs and J. H. B. Christian), Academic Press, London.

Harrigan, W. F. (1983). Personal communication.

Hepner, E. (1980). *Public Health, London,* **94,** 337.

Hill, E. C. (1983). *Laboratory Practice.*

Hill, M. A. (1977). *Proc. Inst. Fd Sci. & Technol.,* **10,** 157.

Hobbs, B. (1965). *National Conference on Salmonellosis,* Public Health Service Publ. 1262, Washington, DC.

Hobbs, B. C. and Gilbert, R. J. (1978). *Food Poisoning and Food Hygiene,* Edward Arnold, London.

Robson, N. C. (1978). *Proc. Inst. Fd Sci. & Technol.,* **11,** 144.

Chapter 8

Laboratory Examination of Frozen Foods

C. A. White and L. P. Hall

*Campden Food Preservation Research Association,
Chipping Campden, UK*

INTRODUCTION

Frozen foods are not sterile products and often contain quite large numbers of micro-organisms, some of which may be potential pathogens. Nevertheless, quick-frozen foods hold an excellent safety record for the consumer, and this must be at least partly attributed to the work of microbiologists in quality control.

The term frozen foods covers a very diverse group of products which are processed in a variety of ways, e.g. some are cooked or blanched prior to freezing, whilst some are frozen in the raw state. Consequently, the sources of contamination and the composition of the microbial flora of these products is very different, and this influences the points at which material should be sampled and what specific microbiological tests should be carried out. However, the methodology for sampling and performing these microbiological tests is common to all types of frozen foods. In this chapter, procedures for sampling frozen foods and appropriate techniques for analysis are recommended, and particular problems discussed.

GENERAL APPROACH TO MICROBIOLOGICAL QUALITY CONTROL

Microbiological tests constitute a useful tool for monitoring standards of factory hygiene, quality of raw materials, and end products.

251

However, effective microbiological quality control depends on a regular sampling programme which provides meaningful information which can then be acted upon.

A system known as hazard analysis, critical control point (HACCP) as described by Bauman (1974) has been proposed as a preventative approach to quality control. Hazard analysis involves a detailed assessment of potential problem areas in food production, such as raw materials, processes and handling procedures, while critical control points are the factors which determine the control of these problem areas.

Production Line Control

Experience will identify hazard points along the production line which may act as a focus for contamination. Ideally, critical control points should be monitored using rapid methods, so that the necessary action can be taken at once, e.g. cleaning contaminated equipment or altering a heat process. However, at present the most reliable data are obtained using conventional plate count procedures, which may require several days incubation before results are available.

Equipment hygiene can be assessed using surface swab and agar contact techniques (Niskanen and Pohja, 1977), but results are often difficult to standardise as variations occur depending on the nature of the site to be tested. Uneven or irregular surfaces, various nozzles and cutting devices have different properties which may affect the adhesion of micro-organisms. Consequently, one of the best methods of assessing hygiene is to sample the food itself during production, as this will show where contaminating organisms are actually introduced onto the food.

Detailed hazard analysis of a factory production line is likely to reveal many potential problem areas. Peterson and Gunnerson (1974) described six critical control points in the relatively simple process of production of blanched frozen vegetables. These occurred at the stages of: washing, blanching, cooling, cutting, inspection and freezing, and could be controlled by attending to equipment sanitation, water quality, personal hygiene and minimising the time delay before freezing.

Microbiological examination procedures can be used to monitor conditions at some of these critical control points, but it is impractical to carry out a great many tests. Continuing the example of production of blanched frozen vegetables, the main critical control points in the process occur after blanching, and suitable points at which samples should be taken for microbiological analysis are (i) at the end of the sorting belt and (ii) directly from the freezer. In practice it is recommended that three 20 g samples should be taken at three times during the day's production, i.e. at the beginning, middle and end.

In the preparation of frozen foods involving combinations of ingredients and separate cooking stages, many critical control points exist. Examples of critical control points in the production of pre-cooked frozen meals include such sites as:

(i) raw materials, particularly where an ingredient may be a potential source of food-poisoning bacteria, such as eggs which may by contaminated with *Salmonella*.

(ii) grinding and mixing stages, where equipment may be difficult to clean and where microbial growth may be aided by increasing the surface area of the substrate, breaking up clumps of cells, and providing aeration.

(iii) after cooking, where time delays occur prior to freezing, particularly when meat is involved as this takes longer to cool than vegetables.

Sample Variation

Microbial infection within a batch of frozen food is not distributed evenly, so that viable counts obtained from a single sample may have little relevance to the bulk from which it came. This uneven distribution of infection is mainly caused by the sporadic nature of contamination on the production line, e.g. food debris may build up on equipment at certain points where it can act as a focus for microbial growth. This material, containing higher numbers of micro-organisms, may subsequently be dislodged by a product surge and re-enter the main flow, thus introducing 'pockets' of infection which are only partially spread throughout the bulk.

This may help to explain the wide variations in viable counts reported for samples of frozen foods (Hall, 1977; Slabyj *et al.*, 1981).

However, Hall (1977) has also demonstrated wide variations in viable counts, particularly on selective media, between samples of a frozen vegetable purée which was uniformly inoculated with a mixed microbial flora prior to freezing. Variations were attributed, in part, to small differences in experimental technique, such as time delay of pouring media into inoculated plates. The effects of such differences were thought to be amplified because of the presence of large numbers of freeze-damaged cells which vary in their sensitivity to environmental conditions (see later).

On some occasions, single samples are of value, when, for example, a problem has arisen and part of the batch has been alleged to have spoiled or caused illness. However, for routine quality control, a standardised multiple sampling procedure should be followed to help compensate for the variability in microbiological counts.

Sampling Plans

The most widely accepted sampling plans are those recommended by the ICMSF (1974), whereby a lot, i.e. a quantity of food produced and handled under uniform conditions, usually within a batch, is accepted or rejected on the basis of microbiological analyses of the required number of sample units.

Results of microbiological examinations are used to divide the sample units into classes. The division between acceptable and marginally acceptable sample units is represented by the symbol m, which may be either a certain level of micro-organisms, or zero in the case of some pathogens. The division between marginally acceptable and unacceptable sample units is known as M, and values lying above this will generally result in rejection of the lot.

Sampling plans can work on a simple two-class or three-class basis. The two-class plan is the simplest, where the decision to accept or reject a lot is based on the number of sample units out of the total number (n) tested which have values greater than m. The maximum allowable number of sample units which can exceed m is known as c. Therefore, when $n = 5$ and $c = 2$, if two or more samples give counts above m the lot will be rejected.

In three-class sampling plans, foods can be divided into acceptable, marginally acceptable, and unacceptable, depending on chosen values of m and M. Values lying between m and M are undesirable, but some may be tolerated. However, any value exceeding M will result in rejection of the whole food lot.

The two-class sampling plan is commonly used when presence/absence tests, such as detection of salmonellae, are applied, whereas three-class plans are useful when recording viable counts. More detailed information on choice of sampling plans can be obtained from ICMSF (1974).

This method of establishing microbiological criteria is basically sound, but its application has at times been criticised. A major source of contention is the choice of values from *m* and *M*. For routine hygiene control of a single production line or factory, the values can be based on the 'norm' indicated by many routine tests carried out, and thus providing a valuable measure of control. However, in many instances trading standards for aerobic plate counts (APC) based on the *m–M* concept have been promulgated, and the values have been calculated from the results of surveys. In these instances, the emphasis is on a mathematical rather than a quality or hazard basis. In deciding values for *m* and *M,* consideration must be given as to whether exceeding these limits indicates bad commercial proctice or a possible loss of quality. Ranges must also be sufficiently wide and flexible to accommodate the very wide variations which occur in results of analyses because of the unique nature of the frozen food substrate, and the effect of low temperatures on the micro-organisms present.

Methods of Sampling

When taking sample units from a food lot, the aim is to obtain data which are representative of the total. Sampling at random is the obvious way of avoiding bias, and this can best be carried out using tables of random numbers (ICMSF, 1974). These tables consist of columns of single digit numbers generated at random, so that each number is completely independent of the one immediately preceding or following it. Each package in the food lot is numbered sequentially, and sample units are chosen according to numbers obtained from the tables. This method of sampling is, however, only possible where the food is in small packages. Where material is stored in bulk, sample units should be taken from widely separated parts of the lot. The number of samples taken depends on the sampling plan, but normally a minimum of five are removed for examination of each lot.

At the time of sampling, as much information as possible about the condition of the food lot should be recorded, including: (a) the temperature of the sample *in situ*, (b) past records of product storage conditions, and (c) examination of the product for evidence of thawing or re-freezing, i.e. excessive ice formation, or dehydration resulting from prolonged storage under conditions of fluctuating temperatures.

Products which are stored in bulk must be sampled aseptically by removing material (at least 200 g) using, for example, a sterile scoop and transferring to a sterile container, such as a screw-capped sample pot or plastic stomacher bag. When the product is in small packages, individual sample units can be removed without disturbing the wrapping. Once selected, the samples of frozen food must not be allowed to thaw prior to analysis, and should be transported to the laboratory in an insulated container surrounded by dry ice, transferred to a cold store, and maintained at a temperature no higher than $-18°C$.

Preparation of the Sample for Microbiological Analysis

Wherever possible, frozen food for microbiological analysis should be sampled directly from the frozen state. This is to minimise variations in microbiological counts arising from:

(i) different rates of thawing, which have been reported to affect the survival of some bacteria (Gebre-Egziabher *et al.*, 1982), and

(ii) delays between thawing and plating which could result in microbial growth.

Sampling directly from the frozen state is simple when the material is free flowing and only requires mixing before aliquots (usually 20 g) are removed aseptically for microbiological analysis. However, when the food is frozen in a solid block, it must be broken up into small pieces, and this may be achieved using a blunt instrument or by tapping on the bench. Sometimes with very large blocks of food, such as meat or fish, it is necessary to use chisels, bone saws, or cartilage knives to remove pieces from the block. When such implements are used, they must be sterilised either by flaming with alcohol or by wrapping and autoclaving.

Alternatively, blocks can be tempered in a refrigerator at below 5°C for a maximum of 12 h to facilitate the removal of material. It must be remembered, however, that slow warming of this type may result in the death of a greater proportion of the microbial population than the rapid thawing which occurs when frozen fragments are put directly into diluent and blended. Haines (1938) established that thawing death of bacteria is more pronounced between $-5°C$ and $-1°C$, and the longer it takes the food to pass through this range the greater will be the death. This effect was confirmed in later work by Gebre-Egziabher *et al.* (1982).

Preparation of the food homogenate

Standard microbiological enumeration procedures for solid food requires the preparation of a food homogenate by blending the sample in diluent using suitable apparatus. In the analysis of frozen food, this procedure has two main functions:

(i) to thaw the sample,
(ii) to break up large particles of food releasing any bacteria embedded within the material, thus giving an even distribution throughout the homogenate.

The food homogenate can be prepared using either a sterile stainless steel blender with electrically driven cutting blades, such as the 'Ato-mix', or a Colworth Stomacher where the sample, which is contained in a sterile plastic bag, is blended by the pummelling action of paddles. The latter method is currently the most popular, since it uses disposable plastic bags and avoids the need to re-sterilise equipment between samples.

Several studies have been carried out to compare the effectiveness of these two techniques for preparation of homogenates of various food types (Sharpe and Jackson, 1972; Emswiler *et al.*, 1977; Thomas and McMeekin, 1980; Thrasher and Richardson, 1980). In general, there were no significant differences between microbiological counts obtained using either technique, although foods with a high fat content (i.e. >20%) sometimes yielded lower counts using the Stomacher compared with the 'Ato-mix'. However, Sharpe and Jackson (1972) showed that increasing the stomaching time gave more comparable results. The recommended stomaching time for most foods is 30 s and 2 min for fatty foods. For frozen food, the longer time of 2 min is recommended to allow for thawing, and for high fat frozen food, stomaching times of up to 5 min should be used.

Difficulties may also arise when the stomacher is used to blend materials containing hard particles such as bone or fruit stones, as these can cause tears in the plastic sample bags. In some cases, this may be overcome using double bags or by inserting a rubber curtain between the sample and the paddles, but even this may not prevent perforations. With such samples, Ato-mix blenders are recommended, although care must be taken to avoid overheating of the sample. Recommended operation times for this apparatus are 1 min at half speed followed by 1 min at full speed.

LABORATORY MEDIA AND METHODS FOR MICROBIOLOGICAL ANALYSIS OF FROZEN FOOD

Sub-Lethal Injury

The major consideration, when selecting media and methods for the isolation and enumeration of micro-organisms from frozen food, is that a high proportion are sub-lethally injured (reversibly injured, damaged, or debilitated). This means that, although they may be capable of proliferation in the food after thawing and can form colonies on a complete, non-selective medium, i.e. one containing all essential nutrients and no inhibitory agents, they cannot grow on certain laboratory media. Ray and Speck (1973a) described two types of sub-lethally injured cells resulting from freezing:

(a) structurally injured: those cells capable of multiplication on a non-selective, complete agar medium, but not on a nutritionally complete medium containing selective agents, e.g. damage to cell permeability barriers allows access of inhibitory agents; and the more severe

(b) metabolically injured: those cells capable of multiplication on a non-selective, complete agar medium, but not on a minimal salts glucose agar medium. This state may be due to an inability to synthesise essential nutrients from simple carbon and nitrogen sources, resulting in an increased requirement for nutrients, such as amino acids or peptides.

The practical implications of sub-lethal injury are that many selective media used for the enumeration of specific groups of micro-organisms are of limited use for direct analysis of frozen food, since a large proportion of the viable population will fail to grow, thus leading to a gross under-estimation of numbers. Ray (1979) has listed many of the selective agents known to be inhibitory to sub-lethally injured cells, and categorised them into: surface active agents, salt and toxic chemicals, antibiotics, dyes and acids. Some selective agents shown to be inhibitory to freeze-damaged cells include bile salts, used for isolation of the enterobacteriaceae (Warseck *et al.*, 1973) and antibiotics for *Clostridium perfringens* (Apiluktivongsa and Walker, 1980).

The problems of using selective media, when dealing with an injured population, can be overcome in two ways: (a) by incorporation of certain ingredients into the selective medium, which stimulate repair and help to overcome the effects of inhibitory agents (Baird-Parker and Davenport, 1965); and (b) incubation of the sample in specialised solid (Powers and Latt, 1979) or liquid (Mossel *et al.*, 1980) resuscitant media capable of restoring vigour to sub-lethally injured organisms prior to exposure to selective agents.

Resuscitation in liquid media prior to plating onto solid selective agar is not very suitable for accurate enumeration procedures, as growth of some cells can occur before repair of injury of the whole population is complete; the approach is best used for presence or absence tests, such as the detection of salmonellae. Liquid resuscitant and selective media can be used for enumeration if a most probable number (MPN) system is used. However, this latter method is subject to such wide variation as to be of questionable value for quantitative analysis (deMann, 1975), so that where methodology is available, plate counts on solid agar are preferred.

Resuscitation in solid media immobilises the cells, so that although uninjured cells divide and grow while others are repairing freeze damage, only one colony is formed from each cell in the initial inoculum. A convenient method is to incubate the sample on a membrane over resuscitant agar prior to transfer to selective media (Holbrook *et al.*, 1980; Mackey *et al.*, 1980). Alternatively the sample can be mixed with molten, cooled resuscitant agar, which is then allowed to set and incubate for a short time, before overpouring with the selective agar (Hall, 1984; Powers and Latt, 1979). The selective agents diffuse into the lower layer during incubation and inhibit growth of sensitive species.

Further examples of methods for resuscitation of sub-lethally injured organisms for selective enumeration or isolation have been reviewed by Mossel and Corry (1977) and Ray (1979).

Diluent

When sampling frozen foods, the choice of diluent for the initial blending and diluting step is important, as it must allow survival of any damaged organisms. Straka and Stokes (1957) demonstrated rapid and extensive destruction of bacteria, during the sampling of frozen pies, when dilutions were carried out in distilled or tap water, and showed that this could be overcome by using 0·1 per cent peptone water for diluting. Hartman and Huntsberger (1961) showed that phosphate buffer diluent gave higher viable counts of micro-organisms from frozen food than distilled water, although 0·1 per cent peptone water was shown to be superior of both of these. Based on this earlier work, Hall (1982) recommends Peptone Buffer Diluent (PBD), containing a weak phosphate buffer at pH 7·0 and 0·1 per cent peptone, for the examination of frozen foods.

ICMSF (1978) recommends 0·1 per cent peptone water or 0·1 per cent peptone + 0·85 per cent NaCl for general use in the blending and dilution of food samples. Whilst 0·1 per cent peptone should provide satisfactory recovery conditions for most micro-organisms in frozen foods, the addition of NaCl is not recommended as this has been reported to reduce recovery of some freeze-damaged cells (Ray and Speck, 1973*a*). Hall (1977) demonstrated 15–20 per cent lower recovery of freeze-damaged *Escherichia coli* when 0·5 per cent NaCl was included in the resuscitation medium, when compared with viable counts on the same medium without NaCl.

PROCEDURE FOR DETERMINING THE MICROBIAL LOAD OF A SAMPLE

The most common method for determination of the level of microbial contamination of frozen food is to perform a mesophilic Aerobic Plate Count (APC) also known as a Standard Plate Count (SPC) at 30°C. Dilutions of the prepared food homogenate are inoculated into an appropriate solid growth medium. After incubation, visible colonies arising from single cells are counted and used to calculate a viable count per gram of the original food sample.

No single plate count procedure can provide conditions capable of supporting growth of all the viable organisms present in a sample of frozen food. However, it is important to select the best medium, method of inoculation, and incubation conditions to obtain the maximum viable count. The following points should, therefore, be considered for the analysis of frozen foods:

Medium. Plate Count Agar (PCA) which contains glucose, tryptone, and yeast extract is currently used throughout the food industry for non-selective enumeration procedures.

Method of inoculation. (i) Pour plate—1 ml of the diluted sample is mixed with sterile, molten, cooled (45°C) agar in a sterile Petri dish, and allowed to solidify. (ii) Spread plate—the inoculum, usually 0·1 ml, is dispensed onto the surface of a pre-dried, solidified agar plate, and spread over the surface of the medium using a sterile glass spreader.

Both of these techniques have advantages and disadvantages, e.g. pre-poured plates required more preparation since the surface must be dried prior to inoculation. Consequently because of greater handling, there is more opportunity for contamination. Also some variations may arise through subtle differences in worker technique. However, plates can be prepared in advance, and this may be more suitable for field sampling. Inoculation using the pour plate method generally uses a larger sample volume (1 ml compared with 0·1 ml for surface inoculation), and theoretically this should give more reproducible results. A disadvantage of the pour plate method is that cells are exposed to hot (45°C) agar, and this may affect the survival of some bacteria. However, comparative studies (Kennedy *et al.*, 1980; Thomas *et al.*, 1981) have shown no significant differences in APCs using either technique, and the choice of method should depend on the facilities and organisation of the individual laboratory.

Incubation temperature. Since small differences in incubation temperature can make a significant difference to the APC, it is important that this is specified and well controlled if meaningful comparisons are to be made between sets of data. ICMSF (1978) recommends an incubation temperature of 30°C as this will allow growth of all mesophiles, i.e. those species which can grow well between 20°C and 45°C with optima of 30–40°C, and also most psychrotrophs, i.e. those species which grow well below 20°C and have optima between 20°C and 30°C.

Mechanisation of the Plate Counting Procedure

Plate counting techniques, although accurate and simple to perform, are laborious, and efforts have been made to automate parts of the procedure. The spiral plate maker developed by Gilchrist *et al.* (1973) is used in many quality control laboratories for surface inoculation of agar plates and is approved by AOAC (1977) for the examination of food and cosmetics.

A small volume of the sample is dispensed onto the surface of a revolving agar plate in a continuously decreasing manner in the form of an Archimedes spiral. After incubation, colonies develop along the line of inoculation, and since the volume deposited on a given area of the plate is known, areas containing discrete colonies can be counted and related to the number of cells in the original inoculum. Unless samples are heavily contaminated, there is no need to dilute further than 1/10, since counts can be determined over a range of $\log_{10}$ $2 \cdot 5$–$6 \cdot 0$ cfu g^{-1} (Kramer *et al.*, 1979), thus eliminating the laborious preparation of dilution series and reducing the number of plates to be inoculated.

Colonies can either be counted manually using a calibrated grid which divides the plate into sectors, or by using an electronic laser counter designed specifically for the spiral plater. The laser counter is capable of greatly reducing the workload by producing automated counts within 5 s. However, careful calibration is required for each food type to obtain accurate readings (Kramer *et al.*, 1979), since food particles in the inoculum may also be counted.

A disadvantage of the spiral plating system is that, with certain food samples, the stylus becomes blocked with particles of food which are sometimes difficult to clear. In addition, the system has limitations for examining samples containing low numbers of micro-organisms where an inoculum of 1 ml in a poured plate offers greater sensitivity than the $0 \cdot 27$ ml dispensed by the spiral plater. However, except in cases where this greater sensitivity is required or where a pour plate method is preferred, the spiral plating system offers substantial savings in labour costs.

Significance of the Aerobic Plate Count

As previously discussed, the determination of the APC of frozen food is useful as a measure of process hygiene when carried out routinely. If high counts occur in the finished product, this can be investigated by frequent sampling at points along the production line to pinpoint infection foci. However, the significance of the APC as a measure of product quality during frozen storage is reduced because, although high counts may indicate that uncontrolled thawing has occurred, low counts do not necessarily mean good quality. This is because the total microbial flora of the product declines during frozen storage, and this death rate is increased at higher storage temperatures up to −7°C (Michener *et al.*, 1960). Consequently, unless there is an accurate time/temperature record of the post-process history of the product, there is no way of knowing whether frozen foods containing low numbers of micro-organisms have been stored (a) for prolonged periods, or (b) above the recommended frozen storage temperature. Since both of these conditions would result in quality loss through physical and chemical changes, the APC of such samples is misleading as a measure of product quality.

Because the APC of frozen food can be reduced by holding under conditions detrimental to the general quality of the product, it is possible that any attempt to impose mandatory standards could be readily abused. In addition, because of the wide variations in microbiological counts which may occur between samples (see above), it is impractical to set rigid standards for the APC of frozen foods.

METHODS FOR ENUMERATION OF SPECIFIC GROUPS OF ORGANISMS

Total Psychrophilic Aerobes

Psychrophiles can be defined as those organisms which grow well at or around 0°C, have an optimum growth temperature of 12–15°C and a maximum of 20°C (Olson and Nottingham, 1980). Because of this low maximum growth temperature psychrophiles are not enumerated as part of an aerobic plate count at 30°C. Generally this is not a problem, but if low temperature abuse of frozen food is suspected, it may be useful to enumerate this group of organisms to determine if spoilage has occurred.

The procedure is essentially the same as for a mesophilic APC except that incubation is carried out at 1–7°C, and plates are examined after 14 and 28 days. Surface inoculation is also recommended, to avoid exposing heat sensitive organisms to agar at 45°C.

Escherichia coli

E. coli is a Gram-negative, non-spore-forming, lactose fermenting rod belonging to the enterobacteriaceae. Since a primary reservoir of *E.*

coli is the gut of man and other warm-blooded animals, the presence of this organism in foods is often taken as evidence of faecal pollution, and the possible presence of related enteric pathogens such as *Salmonella* spp. Whilst this may hold for samples of water and dairy products, it does not necessarily apply to foods in general, since *E. coli* may also be found in habitats such as undisturbed soil (Geldreich *et al.*, 1964) as well as on vegetation and insects (Geldreich *et al.*, 1962).

Raw materials, such as vegetables, may therefore be contaminated with *E. coli* without having any contact with faecal pollution. In addition, it was shown by Miskimin *et al.* (1976) that the occurrence of *E. coli* in raw and prepared foods showed a very poor correlation with the incidence of certain food poisoning organisms, thus limiting its value as an indicator of food safety.

However, because of its heat sensitivity, *E. coli* should not be found on the product following heating and, as such, can be used as an indicator of post-processing contamination of pre-cooked frozen foods and blanched vegetables. This gives a more specific indication of production hygiene that the more widely used tests for 'coliforms' or enterobacteriaceae which, in the case of some products, gives little more information than the total APC.

Freezing and frozen storage greatly reduces viable counts of *E. coli*, and this is influenced by factors such as the freezing substrate, the temperature of storage and the physiological state of the cells prior to freezing (Mackey *et al.*, 1980). Consequently the value of *E. coli* as an indicator of the quality of frozen foods decreases after a lengthy period of frozen storage.

It is recommended that all frozen food should be tested for *E. coli*, both during production and after freezing. A possible exception to this is raw frozen meat and poultry, where *E. coli* is so common as to be of doubtful significance (Goepfert and Kim, 1975; ICMSF, 1974). Limits for *E. coli* in pre-cooked frozen foods as recommended by ICMSF (1974) are $m = 0$, $M = 100$ ($n = 5, c = 2$), but this is quite severe, and perhaps $m = 10$, $M = 500$ ($n = 5, c = 2$) would be more realistic, especially where improved methods, giving greater recovery, are used.

Some strains of *E. coli* are known to cause food poisoning when present in large numbers. However, these enteropathogenic strains, which require specialised techniques for identification, form only a small proportion of the *E. coli* found in nature and exhibit a high degree of animal/human host specificity (Sojka, 1973). Since current techniques for routine analysis of *E. coli* in foods do not distinguish between enteropathogenic and non-enteropathogenic strains (Abbis and Blood, 1982), it is misleading, at this stage, to attach the same significance to *E. coli,* isolated in this way, as to well defined pathogens such as salmonellae.

Isolation and enumeration of *E. coli*

Enumeration of *E. coli* in foods has traditionally been carried out using Most Probable Number (MPN) techniques (ICMSF, 1978). However, these procedures are laborious and time-consuming, normally requiring 5 days for completion when confirmatory tests are included. In addition, inaccuracies arise in enumeration because of the failure of these techniques to detect lactose negative or anaerogenic strains (those producing acid but not gas from lactose), and also because of the large statistical variation in the tables from which the MPN is derived (deMann, 1975). A more serious drawback in the application of these techniques for the examination of frozen foods is that the selective media employed to not support the growth of sub-lethally injured cells. Since approximately 80–90 per cent of the surviving population of *E. coli* may be sub-lethally injured following freezing (Ray and Speck, 1973*b*; Powers and Latt, 1979), a large proportion of the viable population will remain undetected when such media are used.

Anderson and Baird-Parker (1975) have developed a direct plate method (DPM) for the isolation and enumeration of *E. coli* in food which yields results within 24 h. The specificity of this method depends on the production of indole at 44°C when incubated on Tryptone Bile Agar, thus enabling detection of lactose negative and anaerogenic strains. Holbrook *et al.* (1980) have incorporated a short resuscitation step into this method which allows recovery of sub-lethally injured cells, including those damaged by freezing, prior to exposure to selective agents.

This method involves inoculation of the diluted sample onto the surface of a cellulose acetate membrane overlaid on minerals modified glutamate (MMG) agar, a resuscitation medium containing lactose, glutamate, amino acids, and mineral salts. After incubation for 4 h at 37°C, the inoculated membrane is transferred to the surface of a pre-dried plate of TBA, a selective medium containing 0·15% bile salts and tryptone as the sole carbon source, and incubated at 44 ± 0·5°C for a further 18–24 h. Colonies are then tested for indole production using

Vracko–Sherris reagent, which contains 5% *p*-dimethylaminobenz-
aldehyde in N HCl. Indole producing colonies are stained pink within
5 min and, for practical purposes, may be classified as *E. coli* biotype 1.
   A comparative study on enumeration of *E. coli* from raw meat has
been carried out by Rayman *et al.* (1979). This compared the North
American MPN method, i.e. lauryl sulphate tryptose broth at 35–37°C
and 'faecal coliforms' in EC broth at 44·5 ± 0·2°C, with the Anderson
and Baird-Parker DP method. By spreading 1 ml of a 1:5 dilution of
the solid samples onto each membrane, the lower limit of recovery for
the DP method was 2·5 cells per gram. This was comparable with the
sensitivity of the MPN which, in this study, was 3 cells per gram.
   The DP method, which included a resuscitation step for freeze-
damaged cells, was found to be superior to the MPN method for the
analysis of frozen samples, because of its lower variability, rapidity,
and greater recovery of both high and low numbers of cells. Further
studies (Sharpe *et al.,* 1983) have confirmed these findings, and
successfully evaluated this technique for analysis of frozen samples of
beef, Parmesan cheese, and green beans.

Staphylococcus aureus

Staph. aureus is a Gram-positive, non-spore-forming coccus which
causes food poisoning as a result of ingestion of a pre-formed
enterotoxin produced by growth of the organism in food. *Staph. aureus*
itself is heat sensitive, but the enterotoxin is very heat stable (Fung *et
al.,* 1973).

Control of *Staph. aureus* in frozen foods

Raw materials
The hides and skin of many animals are important sources of *Staph.
aureus,* and this frequently leads to contamination of raw meats,
poultry and milk. Most cooking and blanching procedures will result in
the destruction of viable organisms, but not preformed enterotoxin.
However, since *Staph. aureus* generally competes poorly with other
food spoilage micro-organisms, raw meats and poultry are unlikely to
be toxic unless severely spoiled. Exceptions to this are foods containing
fairly high levels of salt, e.g. seafood, which may select for growth of
salt tolerant staphylococci, thus enabling toxic levels to be reached
before obvious spoilage has occurred. Raw and cooked prawns and

shrimps imported from some parts of the world and used as raw materials are frequently contaminated with *Staph. aureus* (Sumner *et al.*, 1982), and should, therefore, be monitored at the raw material stage.

Contamination because of poor hygiene

A major source of *Staph. aureus* is the nasal cavity and sinuses of man, also tne eyes, skin and infected lesions. There is therefore a greater chance of contamination in products subjected to manual handling before freezing, particularly following heat treatment. Since up to 50% of healthy adults are reported to be nasal carriers (Minor and Marth, 1976), it is impractical to eliminate this source of contamination completely. However, the transfer of *Staph. aureus* from humans to food can be greatly reduced by minimising the amount of manual handling particularly at critical stages. Personnel with boils, pustules or infected lesions must be prohibited from coming into contact with food, in accordance with The Food Hygiene General Regulations (1970) Statutory Instrument 1172. Contamination with *Staph. aureus* following heat treatment should therefore be monitored by regularly testing samples of product during the post-heating stages of the production line.

Freezing and frozen storage

Staph. aureus, like many other species of Gram-positive cocci, is particularly resistant to freezing (Georgala and Hurst, 1963; White and Hall, 1984*a*) and moderately resistant to frozen storage (Raj and Liston, 1961). Therefore, freezing cannot be relied upon to give a significant reduction in the level of contamination from *Staph. aureus,* and prolonged frozen storage will not reduce the potential for growth during subsequent thawing (White and Hall, 1984*a*).

Testing for *Staph. aureus*

Because of the widespread risk of contamination from *Staph. aureus* throughout production, it is recommended that all frozen products are tested for this organism. Where foods receive a heat treatment, and where Good Manufacturing Practice is in operation, the incidence of *Staph. aureus* in the final product should be low or absent. In practice, the presence of a few viable cells in occasional samples may be tolerated. However, if frequent samples give counts of *c.* $100\,\mathrm{g}^{-1}$, this warrants investigation of possible sources of contamination.

Enumeration of *Staph. aureus*

Several plating media are available for the enumeration of *Staph. aureus* in foods (ICMSF, 1978), the most popular being Baird-Parker agar (BP: Baird-Parker, 1962). This medium has the advantage of being highly selective for *Staph. aureus*, whilst allowing recovery of damaged cells. The specificity of the medium relies on the ability of *Staph. aureus* (a) to hydrolyse egg yolk lipoprotein and (b) to reduce potassium tellurite, a chemical toxic to many other organisms. After 24–30 h at 37°C, typical colonies are pitch black, shiny, convex and 1·0–1·5 mm diameter with a narrow white margin of precipitate surrounded by a zone of clearing 2–5 mm in width. Plates should be re-examined for typical colonies after a further 24 h incubation. Two selective growth stimulants, glycine and sodium pyruvate, increase the recovery of sub-lethally injured organisms, including those damaged by freezing (Baird-Parker and Davenport, 1965).

Although this is probably the best medium and method for the selective recovery of *Staph. aureus*, some disadvantages remain:

(a) It requires the separate addition of four ingredients to the molten sterilised medium prior to pouring plates, and

(b) once prepared, the plates cannot be stored for more than 24 hours.

A modification was proposed by Holbrook *et al.* (1969) whereby plates prepared without pyruvate could be held for up to one month at 4°C, and the pyruvate added to the surface of each plate (0·5 ml of a 20% sterile solution) prior to drying.

Dehydrated forms of BP media are available from several manufacturers. However, these incorporate the ingredients glycine and pyruvate into the basal medium, which is then autoclaved, and this may affect their ability to act as resuscitants. Collins-Thompson *et al.* (1974) found some commercially prepared dehydrated BP media (BBL; Oxoid) to be superior to five other selective media for enumeration of heat-treated *Staph. aureus*. However, the dehydrated media deteriorated during storage, and this resulted in a much lower recovery (<1%) of heat-damaged cells. Although this effect could be reversed by the addition of fresh sodium pyruvate, unless carefully monitored, the use of such media may lead to inconsistent results.

Another disadvantage of this medium is that the egg yolk clearing reaction is not conclusive for identification of *Staph. aureus*, and a further test is required for confirmation. This involves the demonstration of production of either coagulase or thermostable nuclease (Rayman *et al.*, 1975), both of which require at least a further 24 h incubation time. Nevertheless, BP medium is strongly recommended for the routine examination of frozen foods.

Modifications to Baird-Parker medium, which enable the direct enumeration of coagulase positive *Staph. aureus,* have been proposed. These involve the substitution of egg yolk with pig plasma (Devoyod *et al., 1976*) or pig plasma plus fibrinogen (Hauschild *et al.,* 1979) which result in the production of a halo of precipitated fibrinogen around the coagulase positive colonies.

Another simple modification of Baird-Parker medium, incorporating the thermostable nuclease test, has also been proposed (Lachia, 1980). This involves heating incubated plates of Baird-Parker medium to inactivate heat labile nucleases in the bacterial colonies, followed by overpouring with toluidine blue deoxyribonucleic acid agar. After further incubation (3 h at 37°C), colonies with bright pink halos are confirmed as *Staph. aureus.* This method has the advantage of yielding results within a few hours.

Salmonella

Food poisoning caused by *Salmonella* spp. is the result of ingestion of significant numbers of viable cells, which then infect and give rise to a mild or severe gastroenteritis depending on the serotype and/or health of the individual. Salmonellae are Gram-negative, non-spore-forming rods which are sensitive to environmental extremes, but which grow rapidly under favourable conditions.

Control of salmonellae in frozen foods

Sources of contamination
The main reservoir for salmonellae is the gut and viscera of warm-blooded animals (Baird-Parker, 1980). Consequently meats, and in particular poultry and eggs, may become contaminated, and where these are used as ingredients, the extent of infection should be monitored and steps taken to prevent its spread through the factory.

Heat treatment
Normal cooking or pasteurisation destroys salmonellae, and this is a major control point in the manufacture of pre-cooked frozen foods.

However, poor hygiene within the factory may result in cross-contamination from raw materials onto the processed product, and this must be prevented by such measures as physical separation of cooked and raw food, avoiding transfer of utensils or containers from one area to the other, and restricting movements of factory personnel (Silliker, 1982). The examination of product, from the post-cooking areas, for salmonellae will assess the effectiveness of these measures.

Freezing and frozen storage

Although salmonellae, in common with other Gram-negative rods, are highly sensitive to freezing and thawing, a proportion of cells will survive (Olson *et al.*, 1981). Many of these survivors will persist during frozen storage and remain capable of growth in the product during thawing (White and Hall, 1984a).

Testing for *Salmonella*

The incidence of *Salmonella* spp. in vegetables is low, and examination of these products is only recommended where there is a possibility of contamination from polluted irrigation water or fertilisers based on sewage. Testing for salmonellae is, however, routine for all raw and pre-cooked frozen foods containing meat, poultry, crustacea, and fresh water fish. It is also recommended that some frozen products containing eggs and milk should be examined for salmonellae.

Salmonellae should not be present in any 25 g sample of frozen food, but although this standard can be attained with respect to pre-cooked foods by strict control of hygiene, particularly following heat treatment, it is difficult to eliminate salmonellae from frozen foods which receive no heat treatment. The simple answer would be to use raw materials which are not contaminated with salmonellae, but where meat, and particularly poultry, is concerned this is, at present, impracticable (Baird-Parker, 1980). The frozen food producer must, therefore, be aware of the problem and keep the level of contamination to an absolute minimum using appropriate control measures.

Detection of salmonellae

Methods for detection of salmonellae in frozen foods must enable detection of low numbers, often in the presence of much larger numbers of associated micro-organisms. Since the presence of any salmonellae is of concern, enrichment is used and results are expressed as presence or absence in a given weight of sample. The principle steps in the isolation procedure are:

1. Pre-enrichment in non-selective broth.
2. Selective enrichment in specialised broth.
3. Isolation on selective solid media.
4. Confirmation by biochemical and serological tests.

A wide variety of selective media and analytical schemes are available for detection of salmonellae in foods, and there is much debate as to which procedure gives the highest rate of isolation. It is important that the best method of isolation should be used for each food type, and also that procedures should be standardised so that meaningful comparisons can be made between laboratories.

1. With respect to isolation of salmonellae from frozen food, pre-enrichment in non-selective media is the most important stage, since this allows damaged cells to resuscitate prior to exposure to selective agents in subsequent enrichment broths. Several types of pre-enrichment media have been proposed for different foods, but Schothorst *et al.* (1978) found no significant differences using tryptone soya broth, glucose mineral salts medium and buffered peptone water for isolation of *Salmonella* spp. from frozen meat; Gröhn and Hirn (1982) found 1% peptone water satisfactory for pre-enrichment of salmonellae from frozen chicken. On the other hand, the composition of the competitive flora can have a major influence on the resuscitation and growth of *Salmonella* spp. during enrichment, particularly where, for example, lactic acid bacteria grow and reduce the pH (Sadovski, 1977). Buffered peptone water is, therefore, recommended, as this helps to keep the pH within tolerable limits for the resuscitation and growth of *Salmonella* spp. The addition of excess calcium carbonate is also very effective in stabilising the pH of the pre-enrichment mixture (S. A. Rose, Pers. comm.), and this has been used to improve the isolation of salmonellae from acid products.

2. Once the injured salmonellae in a frozen sample are restored to full vigour, the problems of isolation are common to other food types. The success of one selective enrichment broth over another following a pre-enrichment step depends, therefore, more on serotype and composition of the competitive flora than on the nature of the original preservative treatment.

The most popular selective enrichment media for salmonellae are probably selenite cystine and Muller–Kauffmann tetrathionate brilliant green broth as recommended by ICMSF (1978). Incubation of these media at 43°C has been shown (Silliker and Gabis, 1974) to improve results compared with incubation at 35°C, presumably because of selective inhibition of the competitive flora. Some analytical schemes use both media in parallel, thus increasing the chances of isolation as the options will favour a wider range of serotypes.

Another popular selective enrichment broth is Rappaport's magnesium chloride malachite green medium which has been modified several times (Vassiliadis, 1983). This medium was evaluated by Harvey and Price (1981), and found to be more efficient than Muller–Kauffmann tetrathionate and selenite F broth for isolation of salmonellae from chicken giblets following pre-enrichment. Similarly Vassiliadis *et al.* (1981) found Rappaport's medium to be superior to Muller–Kauffmann tetrathionate broth for isolation of salmonellae from pork sausages. The improved isolation rate obtained with both versions of Rappaport's broth was attributed to its strong inhibitory effect on competing micro-organisms.

3. After the enrichment stages, cultures are streaked onto selective agar to isolate typical colonies, which can then be taken for confirmation. The most common selective agar is brilliant green phenol red agar (BGA), and this is highly recommended. Usually a second isolation agar is used in conjunction with BGA, again to cover a wider range of serotypes, these include: bismuth sulphite, desoxycholate citrate, desoxycholate citrate lactose sucrose, and xyline lysine desoxycholate agars, all of which are available commercially from Oxoid Ltd.

4. Typical colonies from any of the above media still need to be confirmed as *Salmonella* spp. This is usually achieved by inoculating into confirmatory tubes, such as Kohn's two tube media plus lysine broth or triple sugar iron agar plus lysine iron agar. If isolates produce typical 'salmonella' reactions, they can be confirmed by testing for polyvalent O (somatic) and polyvalent H (flagellar) antigens by agglutination reactions with the appropriate antisera.

Isolates can be classified further as serotypes, but this is normally only carried out in specialised laboratories. For the purposes of routine examination of food samples, it is sufficient to identify isolates as *Salmonella* spp.

Clostridium perfringens

Cl. perfringens is a Gram-positive, anaerobic, spore-forming rod, which causes food poisoning as a result of ingestion of large numbers of vegetative cells which release an enterotoxin during sporulation within the intestine.

Control of *Cl. perfringens* in frozen foods

Sources of contamination
Raw materials such as meat, poultry, herbs and spices should be monitored as they may be heavily contaminated.

Heat treatment
Normal cooking of meat and poultry will destroy vegetative cells, but may allow survival of the spores and stimulate their germination (Roberts, 1972). Since growth of *Cl. perfringens* can be very rapid (generation time, 20 min at 33–49°C) the product must be cooled quickly to below 15°C where growth is either very slow or absent (Craven, 1980), and frozen with the minimum delay to prevent excessive multiplication of any survivors.

Freezing and frozen storage
Vegetative cells of *Cl. perfringens* are sensitive to freezing and frozen storage, but the spores are very resistant (Strong and Canada, 1964; Trakulchang and Kraft, 1977). However, both spores and vegetative cells can survive prolonged frozen storage and remain capable of growth in the product during thawing (White and Hall, 1984*b*).

Testing for *Cl. perfringens*

Because of its association with meat and poultry products, it is recommended that all pre-cooked frozen foods containing these ingredients should be routinely tested for *Cl. perfringens*.

Even with strict control measures, *Cl. perfringens* may still be found in low numbers in some frozen foods, and a low incidence may be tolerated. However, if frequent samples contain $100 \, \text{cfu g}^{-1}$, this warrants investigation of raw materials and processing conditions.

The examination of foods for *Cl. perfringens* follows a two-step procedure:

1. enumeration of presumptive *Cl. perfringe,* incubation at 37°C on a primary selective me(
2. confirmation of isolates by further tests.

Presumptive enumeration—Several selective me(analysis of foods for *Cl. perfringens*, most of formation of black colonies as a result of sulphite reduction in the presence of an iron salt, and incorporation of one or more antibiotics for suppression of interfering organisms.

Harmon *et al.* (1971) used D-cycloserine, sodium metabisulphite, and egg yolk as the basis of selectivity in tryptose–sulphite–cycloserine (TSCE) agar. The egg-yolk free modification of TSC proposed by Hauschild and Hilsheimer (1974*a*) gave similar recovery and selectivity, but had the advantages of being easier to prepare and allowing enumeration of strains having low lecithinase activity, which do not give a good egg yolk reaction and might otherwise be overlooked. This was the medium of choice for enumeration of *Cl. perfringens* in foods as a result of a comparative study carried out on behalf of ICMSF (Hauschild *et al.*, 1977).

In the analysis of frozen foods for *Cl. perfringens*, recovery of freeze-damaged cells must be considered, since increased sensitivity to selective agents might be expected. Apiluktivongsa and Walker (1980) compared a wide range of antibiotics, and found D-cycloserine to be the least inhibitory to cold-stressed (frozen and chilled) cells of *Cl. perfringens*, particularly when in combination with sodium metabisulphite. This supports the findings of Hall (1980) and Orth (1977) where TSC agar, containing these ingredients, was shown to be superior to oleandomycin–polymyxin–sulphadiazine perfringens (OPSP) medium and sulphite polymyxin sulphadiazine (SPS) medium, respectively, for recovery of *Cl. perfringens* following freezing. However, the main advantage of TSC over OPSP agar is its greater selectivity against interfering spoilage organisms (Hauschild and Hilsheimer, 1974*b*; Adams and Mead, 1980; Hall, 1980) which usually enables confirmation of typical colonies to be carried out without a purification step.

For these reasons, TSC medium is strongly recommended as a primary selective medium in the analysis of frozen foods for *Cl. perfringens*.

Confirmation

Until a more suitable selective medium is developed for *Cl. perfringens* it is still necessary to confirm presumptive colonies by further tests. The procedure adopted by the AOAC (Harmon, 1976) for confirmation of isolates on TSC agar was found to be simple to perform and the results easy to read (Hall, 1980). This requires only two media; buffered motility nitrate (BMN) agar and lactose gelatin medium (LGM) to test for nitrate reduction, non-motility, lactose fermentation, and gelatin liquefaction. Enrichment of isolates from the primary selective medium may be necessary to obtain a good inoculum for the confirmatory tests, and this can be achieved by incubation of black colonies in fluid thioglycolate medium (FTM). If typical and atypical colonies are crowded on the agar (as is often the case with OPSP agar), isolates need to be purified before confirmation. This is done by streaking onto selective agar to obtain single, well isolated colonies which can be used directly or enriched in FTM.

Bacillus cereus

B. cereus is a Gram-positive, aerobic or facultatively anaerobic, spore-forming rod which is known to cause two distinct types of food poisoning syndromes resulting in either vomiting or diarrhoea (Gilbert and Parry, 1977). The vomiting-type syndrome is almost exclusively associated with cooked rice contaminated with a heat-resistant enterotoxin produced during growth of *B. cereus*, while the diarrhoeal syndrome is associated with a wide variety of foods including prepared meat dishes and dairy products (Hutchinson and Taplin, 1978).

Control of *B. cereus* in frozen foods

Sources of contamination
The spores are common in soil, and a wide variety of foods and food ingredients are contaminated with this organism, particularly dried foods such as rice, pulses, cereals (Blakely and Priest, 1980), herbs and spices, but also dairy products and meat.

Heat treatment
Like *Clostridium perfringens,* the spores of *Bacillus cereus* are heat resistant and will survive light cooking (Parry and Gilbert, 1980). Conditions following heat treatment must, therefore, minimise the opportunity for growth to occur prior to freezing. This can be achieved by adequate cooling (to <10°C), and avoiding stoppages on the production line.

Freezing and frozen storage
The vegetative cells of *B. cereus* are sensitive to freezing and frozen storage, but the spores are resistant and remain viable at least for six months at −18°C (White and Hall, 1984*b*).

Testing for B. cereus

It is recommended that all frozen foods containing meat, dairy products and, especially, cereals should be routinely examined for *B. cereus*. Low numbers of vegetative cells and spores may be tolerated in some samples, but viable counts exceeding 1000 g^{-1} warrant investigation of sources of contamination and processing conditions.

Isolation and enumeration of B. cereus

A number of selective media are available for the selective enumeration of *B. cereus* in foods. Mannitol egg yolk polymyxin (MYP) agar, as described by Mossel *et al.* (1967), relies on the inability of *B. cereus* both to dissimilate mannitol and to give a positive egg yolk precipitin reaction. This medium is recommended by the ICMSF (1978) and the AOAC (Lancett and Harmon, 1980) for the routine examination of foods for *B. cereus*. However, Holbrook and Anderson (1980) have described an improved medium, polymyxin pyruvate egg yolk mannitol bromothymol blue agar (PEMBA), based on similar characteristics, but claimed to be more selective against other micro-organisms in foods and more diagnostic for *B. cereus*. The quantitative enumeration of *B. cereus* from a variety of foods using this medium was found to be comparable with three other media, including MYP (Holbrook and Anderson, 1980).

Unfortunately none of these media is completely selective for *B. cereus*, and while they may be sufficient for routine screening of food samples, further identification of presumptive isolates may be required. Confirmatory tests following isolation on PEMBA (Holbrook and Anderson, 1980) are more rapid than those described by Mossel *et al.* (1967) for confirmation of isolates on MYP. PEMBA is, therefore, recommended because of its greater selectivity when dealing with low numbers of *B. cereus* in foods.

Vibrio parahaemolyticus

V. parahaemolyticus is a Gram-negative, halophilic, facultatively anaerobic rod which causes a food poisoning infection usually as a result of consumption of contaminated seafoods. It is the major cause of food poisoning in Japan, where it is customary to eat raw or semi-processed fish, although it is also associated with outbreaks involving cooked seafoods (Desmarchelier, 1978).

Control of *V. parahaemolyticus* in frozen foods

Sources of contamination

The primary source of *V. parahaemolyticus* is the marine environment, particularly coastal and estuarine waters of South East Asia and Japan. Consequently, seafood, including marine fish, crustaceans and molluscs, may frequently be contaminated with this organism (Sakazaki, 1973).

Processing conditions

Light cooking, or other mild heat treatments are sufficient to destroy this organism (Lee, 1973). However, growth rates of *V. parahaemolyticus* are very high under suitable conditions, and care must be taken, on the production line, to avoid hold-up of the materials in the warm state. Also, when the materials are cooked before freezing, or form ingredients of pre-cooked dishes, precautions must be taken to eliminate cross-contamination from raw to cooked products.

Freezing and frozen storage

As with most Gram-negative organisms, the vibrios are highly sensitive to freezing, but although the levels of *V. parahaemolyticus* will be reduced by freezing and frozen storage (Covert and Woodburn, 1972; Ray *et al.*, 1978), some survivors may persist and remain viable in the frozen product (Lee, 1973).

Testing for *V. parahaemolyticus*

This is recommended when seafoods from suspect sources are to be frozen and consumed without cooking. The limit suggested by the ICMSF (1974) is 100 g^{-1}.

Isolation and enumeration of *V. parahaemolyticus*

V. parahaemolyticus has an obligate requirement for salt, and rapidly loses viability when added to water (Lee, 1972). All dilution fluids and growth media must, therefore, contain an adequate concentration of salt to maintain viability and permit growth. Although *V. parahaemolyticus* can grow at salt concentrations from 0·5 to 7 per cent (ICMSF, 1978), working levels are normally between 0·5 and 3 per cent.

As with salmonellae, procedures for detection of *V. parahaemolyti-*

cus in foods usually employ liquid enrichment prior to isolation on solid media. Selective enrichment broths in current use include: glucose salt teepol broth (GSTB), salt colistin broth (SCB), salt polymyxin B broth (SPB), and thiosulphate citrate bile salt sucrose broth (TCBS) (ICMSF, 1978; Hall, 1982; Taylor *et al.*, 1982). However, Ray *et al.* (1978) showed freeze-damaged cells of *V. parahaemolyticus* to be sensitive to some of the ingredients in these enrichment broths, including NaCl at 3 per cent. Pre-enrichment, initially in TSB + 0·5 per cent NaCl at 35°C followed by overnight culture in TSB + 3 per cent NaCl was, therefore, recommended to allow for repair prior to transferring to selective broths.

Samples incubated in one or more selective enrichment broths are then subcultured by streaking onto selective agar to obtain single colonies. The most widely used selective medium is thiosulphate citrate bile salt sucrose (TCBS, Oxoid), which contains bile salts, deoxycholate and sodium thiosulphate as selective agents, with bromothymol blue in combination with thymol blue as an indicator system for sucrose fermentation. Most strains of *V. parahaemolyticus* are sucrose negative, and characteristic colonies after 24 h at 37°C on TCBS agar are colourless with a green centre. However, some strains are sucrose positive, resulting in yellow colonies similar to *Vibrio cholera* or *Vibrio alginolyticus*. In both cases, presumptive isolates of *V. parahaemolyticus* need to be confirmed by additional tests, such as those described by ICMSF (1978), Hall (1982) or Taylor *et al.* (1982).

Enumeration of *V. parahaemolyticus* is sometimes carried out by direct plating on to TCBS agar, but this will not support growth of freeze-damaged cells, and *c.* 80 per cent of the viable population will not be detected (Ray, 1979). A most probable number (MPN) procedure has been suggested by Ray *et al.* (1978) using the same pre-enrichment method as described for the isolation procedure, and GSTB as the selective broth. This repair-detection method was shown to be superior to the MPN procedure recommended by ICMSF (1978) for enumeration of *V. parahaemolyticus* from frozen seafoods.

Rapid Methods

All the methods described so far for the microbiological analysis of frozen foods rely on colony counting techniques, but these require at least 24 h incubation to enable the development of visible colonies. An alternative approach is to measure some aspect or effect of the metabolic activity of the micro-organisms, which can then be related to the level of microbial contamination in the food sample.

Measurement of adenine triphosphate (ATP) using bioluminescence

The high energy molecule ATP is found in all living cells, including bacteria. By measuring the total amount of ATP in a bacterial culture, and knowing the average ATP content of a single cell, it is possible to calculate the number of bacteria in the culture.

The most sensitive method of measuring ATP is by using the bioluminescence reaction of the firefly, *Photinus pyralis*. When ATP is mixed with luciferin and luciferase from the firefly in the presence of oxygen, light is emitted in proportion to the concentration of ATP in the mixture. By comparison with standards, the concentration of ATP in the sample can be determined. Instruments are available (Lumac BV, LKB Instruments) which use this reaction for ATP determinations.

However, although this technique is straightforward when applied to the analysis of bacterial cultures, in food samples, the intrinsic ATP in the food itself must first be destroyed before measuring the level of bacterial ATP. This can be achieved either by selective release and removal of non-bacterial ATP with a surfactant and ATP-hydrolysing enzyme prior to releasing ATP from the bacterial cells (Chappelle *et al.*, 1978), or alternatively by physical separation of the bacteria, using for example filtration or centrifugation (Stannard and Wood, 1984). Further studies are required to evaluate this technique for the analysis of frozen foods, where the average ATP content per cell, which is known to be influenced by growth rates (Cole *et al.*, 1967) may differ significantly from that in non-frozen samples.

A highly promising application of this technique may be in the monitoring of line hygiene, either by product or surface sampling, since results can be obtained within 1 h, thus allowing any corrective action to be taken before more serious problems develop.

Measurement of electrical impedance

When micro-organisms grow, the chemical composition of the substrate is altered, i.e. nutrients are consumed and metabolic end-products released. This is associated with a change in impedance, i.e. the resistance to flow of an electrical current, of the growth medium. This change only occurs when microbial growth reaches a certain level (10^6–10^7 cfu ml^{-1}) and, when conditions are standardised, the time taken to reach this threshold can be related to the initial concentration of micro-organisms in the sample.

Hardy *et al.* (1977) have used this technique for assessing the acceptability ($<10^5$ cfu g^{-1}) or unacceptability ($>10^5$ cfu g^{-1}) of frozen vegetables. Depending on the initial level of contamination, analysis was completed within 5 h, and there was good agreement ($>92\%$) with standard plate count procedures.

Equipment is available for measuring impedance (Bactomatic Inc., Malthus Instruments) which allows a large number of samples to be analysed simultaneously, and generally gives results within a working day.

Both the ATP monitoring system and the impedimetric techniques show promise for the replacement of the APC in certain areas of microbiological quality control of frozen foods. However, a large proportion of microbiological quality control entails the detection and enumeration of specific organisms using selective and diagnostic media. Efforts have been made with impedance measurements to estimate numbers of 'coliforms' by incubation of samples in selective media (Martins and Selby, 1980), but analysis time was increased to 24 h, the same as a plate count on violet red bile agar.

It would appear that, at least in the short term, rapid methods based on ATP or impedimetric measurements are limited to assessing the total microbial load of a sample, and that the rapid detection of specific groups of organisms may require a completely different approach. An example of an alternative approach to traditional selective techniques is the use of DNA–DNA hybridisation for the detection of salmonellae in foods (Fitts *et al.*, 1983). Although still in the experimental stage, this technique offers greater specificity and is more rapid than existing methods, which are laborious and have persistent problems of false positive and false negative reactions.

REFERENCES

Abbis, J. S. and Blood, R. M. (1982). In: *Isolation and Identification Methods for Food Poisoning Organisms* (Eds J. E. L. Corry, D. Roberts and F. A. Skinner), Academic Press, London.

Adams, B. W. and Mead, G. C. (1980). *Journal of Hygiene, Cambridge*, **84**, 151–8.

Anderson, J. M. and Baird-Parker, A. C. (1975). *Journal of Applied Bacteriology*, **39**, 111–17.

AOAC (1977). *Journal of the Association of Official Analytical Chemists*, **60**, 493–4.

Apiluktivongsa, P. and Walker, H. W. (1980). *Journal of Food Science*, **45**, 574–6.

Baird-Parker, A. C. (1962). *Journal of Applied Bacteriology*, **25**, 12–19.

Baird-Parker, A. C. (1980). *Food Technology in Australia*, **32**(5), 254–60.

Baird-Parker, A. C. and Davenport, E. (1965). *Journal of Applied Bacteriology*, **28**, 390–402.

Bauman, H. E. (1974). *Food Technology*, **28**(9), 30–4.

Blakely, L. J. and Priest, F. G. (1980). *Journal of Applied Bacteriology*, **48**, 297–302.

Chappelle, E. W., Piccolo, G. L. and Deming, J. W. (1978). *Methods in Enzymology*, **57**, 65–72.

Cole, H. A., Wimpenny, J. W. T. and Hughs, D. E. (1967). *Biochimica Biophysica Acta*, **143**, 445–53.

Collins-Thompson, D. L., Hurst, A. and Aris, B. (1974). *Canadian Journal of Microbiology*, **20**, 1072–5.

Covert, D. and Woodburn, M. (1972). *Applied Microbiology*, **23**(2), 321–5.

Craven, S. E. (1980). *Food Technology*, **34**(4), 80–7.

Desmarchelier, P. (1978). *Food Technology in Australia*, **30**(9), 339–45.

Devoyod, J. J., Millet, L. and Moquot, G. (1976). *Canadian Journal of Microbiology*, **22**(11), 1603–11.

Emswiler, B. S., Pierson, C. J. and Kotula, A. W. (1977). *Food Technology*, **31**, 40–2.

Fitts, R., Diamond, M., Hamilton, C. and Neri, M. (1983). *Applied Environmental Microbiology*, **46**(5), 1146–51.

Fung, D. Y. C., Steinberg, D. H., Miller, R. D., Kurantnick, M. J. and Murphy, T. F. (1973). *Applied Microbiology*, **26**, 938–42.

Gebre-Egziabher, A., Thomson, B. and Blankenagel, G. (1982). *Journal of Food Protection*, **45**(2), 125–6.

Geldreich, E. E., Huff, C. B., Bordner, R. H., Kabler, P. W. and Clark, H. F. (1962). *Journal of Applied Bacteriology*, **25**, 87–93.

Geldreich, E. E., Kenner, B. A. and Kabler, P. W. (1964). *Applied Microbiology*, **12**(1), 63–9.

Georgala, D. L. and Hurst, A. (1963). *Journal of Applied Bacteriology*, **26**, 346–58.

Gilbert, R. J. and Parry, J. M. (1977). *Journal of Hygiene, Cambridge*, **78**, 69–74.

Gilchrist, J. E., Campbell, J. E., Donnelly, G. B., Peeler, J. T. and Delaney, J. M. (1973). *Applied Microbiology*, **25**, 244–52.

Goepfert, J. M. and Kim, H. V. (1975). *Journal of Milk and Food Technology*, **38**, 449–52.

Gröhn, K. and Hirn, J. (1982). *Suomen Eläinlääkärilehti*, **88**(3), 145–8.

Haines, R. B. (1938). *Proceedings of the Royal Society*, **B124**, 451–63.

Hall, L. P. (1977). Technical Memorandum No. 182, Campden Food Preservation Research Association, Chipping Campden, Glos., UK.

282   C. A. White and L. P. Hall

Hall, L. P. (1980). Technical Memorandum No. 238, Campden Food Preservation Research Association, Chipping Campden, Glos., UK.

Hall, L. P. (1982). *A Manual of Methods for the Bacteriological Examination of Frozen Foods*, 3rd Edn, Campden Food Preservation Research Association, Chipping Campden, Glos., UK.

Hall, L. P. (1984). *Journal of Applied Bacteriology*, **56**, 227–35.

Hardy, D., Kraeger, S. J., Dufour, S. W. and Cady, P. (1977). *Applied Environmental Microbiology*, **84**(1), 14–17.

Harmon, S. M. (1976). *Journal of the Association of Official Analytical Chemists*, **59**, 606–12.

Harmon, S. M., Kautter, D. A. and Peeler, J. T. (1971). *Applied Microbiology*, **22**(4), 688–92.

Hartman, P. A. and Huntsberger, D. V. (1961). *Applied Microbiology*, **9**, 32.

Harvey, R. W. S. and Price, T. H. (1981). *Journal of Hygiene, Cambridge*, **87**(2), 219–24.

Hauschild, A. H. W. and Hilsheimer, R. (1974a). *Applied Microbiology*, **27**(3), 521–6.

Hauschild, A. H. W. and Hilsheimer, R. (1974b). *Applied Microbiology*, **27**(1), 78–82.

Hauschild, A. H. W., Gilbert, R. J., Harmon, S. M., O'Keefe, M. F. and Vahlefeld, R. (1977). *Canadian Journal of Microbiology*, **23**, 884–92.

Hauschild, A. H. W., Park, C. E. and Hilsheimer, R. (1979). *Canadian Journal of Microbiology*, **25**, 1052–7.

Holbrook, R. and Anderson, J. M. (1980). *Canadian Journal of Microbiology*, **26**, 753–9.

Holbrook, R., Anderson, J. M. and Baird-Parker, A. C. (1969). *Journal of Applied Bacteriology*, **32**, 187–92.

Holbrook, R., Anderson, J. M. and Baird-Parker, A. C. (1980). *Food Technology in Australia*, **32**(2), 78–83.

Hutchinson, E. M. S. and Taplin, J. (1978). *Food Technology in Australia*, **30**, 329–33.

ICMSF (1974). *Microoganisms in Foods, 2. Sampling for Microbiological Analysis: Principles and Specific Applications*. Publ. for the International Commission on Microbiological Specifications for Foods by University of Toronto Press.

ICMSF (1978). *Microorganisms in Foods, 1. Their Significance and Methods of Enumeration*. Publ. for the International Commission on Microbiological Specifications for Foods by University of Toronto Press.

Kennedy, J. E. Jr, Philips, P. E. and Oblinger, J. L. (1980). *Journal of Food Protection*, **43**, 592–4, 600.

Kramer, J. M., Kendal, M. and Gilbert, R. J. (1979). *European Journal of Applied Microbiology and Biotechnology*, **6**, 289–99.

Lachia, R. V. (1980). *Applied Environmental Microbiology*, **39**(1), 17–19.

Lancett, G. A. and Harmon, S. M. (1980). *Journal of the Association of Official Analytical Chemists*, **63**, 581–6.

Lee, J. S. (1972). *Applied Microbiology*, **23**, 166–7.

Lee, J. S. (1973). *Journal of Milk and Food Technology*, **36**(8), 405–8.

Mackey, B. M., Derrick, C. M. and Thomas, J. A. (1980). *Journal of Applied Bacteriology*, **48**, 318–24.

deMann, J. C. (1975). *European Journal of Applied Microbiology*, **1**, 67–78.

Martins, S. B. and Selby, M. J. (1980). *Applied Environmental Microbiology*, **39**, 518–24.

Michener, H., Thompson, P. A. and Dietrich, W. C. (1960). *Food Technology*, **14**, 290–4.

Minor, T. E. and Marth, E. H. (1976). In: *Staphylococci and their Significance in Foods*, Chapter 9, Elsevier Science Publishers, Amsterdam.

Miskimin, D. K., Berkowitz, K. A., Solbert, M., Rihajr, W. E., Franke, W. C., Buchanan, R. L. and O'Leary, V. (1976). *Journal of Food Science*, **41**, 1001–6.

Mossel, D. A. A. and Corry, J. E. L. (1977). *Alimenta Zurich*, **16**, Special Issue on Microbiology, 19–34.

Mossel, D. A. A., Koopman, M. J. and Jongerius, E. (1967). *Applied Microbiology*, **15**, 650–3.

Mossel, D. A. A., Veldman, A. and Eelderink, I. (1980). *Journal of Applied Bacteriology*, **49**, 405–19.

Niskanen, A. and Pohja, M. S. (1977). *Journal of Applied Bacteriology*, **42**, 53–63.

Olson, J. C. Jr, and Nottingham, P. M. (1980). In: *Microbial Ecology of Foods, 1. Factors Affecting Life and Death of Microorganisms*. Chapter 1. Publ. for The International Commission on Microbiological Specifications for Foods by Academic Press, London.

Olson, V., Swaminathan, B. and Stadelman, W. J. (1981). *Journal of Food Science*, **46**(5), 1323–6.

Orth, D. S. (1977). *Applied Environmental Microbiology*, **33**(4), 986–8.

Parry, J. M. and Gilbert, R. J. (1980). *Journal of Hygiene, Cambridge*, **84**, 77–82.

Peterson, A. C. and Gunnerson, R. E. (1974). *Food Technology*, **28**(9), 37–44.

Powers, E. M. and Latt, T. G. (1979). *Journal of Food Protection*, **42**(4), 342–5.

Raj, H. and Liston, J. (1961). *Food Technology*, **15**, 429–34.

Ray, B. (1979). *Journal of Food Protection*, **42**(4), 346–55.

Ray, B. and Speck, M. L. (1973*a*). *CRC Critical Reviews in Clincial Laboratory Sciences*, **4**(2), 161–213.

Ray, B. and Speck, M. L. (1973*b*). *Applied Microbiology*, **25**(4), 499–503.

Ray, B., Hawkins, S. M. and Hackney, C. R. (1978). *Applied Environmental Microbiology*, **35**(6), 1121–7.

Rayman, M. K., Park, C. E., Philpott, J. and Todd, E. C. D. (1975). *Applied Microbiology*, **29**, 451–4.

Rayman, M. K., Jarvis, G. A., Davidson, C. M., Long, S., Allen, J. M., Tong, T., Dodsworth, P., McLaughlin, S., Greenberg, S., Shaw, B. G., Beckers, H. J., Qvist, S., Nottingham, P. M. and Stewart, B. J. (1979). *Canadian Journal of Microbiology*, **25**, 1321–7.

Roberts, D. (1972). *Journal of Hygiene, Cambridge*, **70**, 565–88.

Sadovski, A. Y. (1977). *Journal of Food Technology*, **12**, 85–91.

Sakazaki, R. (1973). In: *Microbiological Safety of Foods* (Eds B. C. Hobbs and J. H. B. Christian), pp. 375–85, Academic Press, London.

Schothorst, M. Van, Gilbert, R. J., Harvey, R. W. S., Pietzsch, O. and Kampelmacher, E. H. (1978). *Zentralblatt für Bacteriologie, Parasitenkunde, Infectionskrankheiten und Hygiene*, I Abt Orig, **B 167,** 138–45.

Sharpe, A. N. and Jackson, A. K. (1972). *Applied Microbiology*, **24**(2), 175–8.

Sharpe, A. N., Rayman, M. K., Burgener, D. M., Conley, D., Loit, A., Milling, M., Peterkin, P. I., Purvis, U. and Malcolm, S. (1983). *Canadian Journal of Microbiology*, **29**, 1247–52.

Silliker, J. H. (1982). *Journal of Food Protection*, **43**, 307–13.

Silliker, J. H. and Gabis, D. A. (1974). *Canadian Journal of Microbiology*, **20**, 813–16.

Slabyj, B. M., Martin, R. E. and Ramsdell, G. E. (1981). *Journal of Food Science*, **46**(3), 716–9.

Sojka, W. J. (1973). *Canadian Institute of Food Science and Technology*, **6**(2), 52–63.

Stannard, C. J. and Wood, J. M. (1984). *Journal of Applied Bacteriology*, **55**(3), 429–39.

Straka, R. P. and Stokes, J. L. (1957). *Applied Microbiology*, **5**, 21–5.

Strong, D. H. and Canada, J. C. (1964). *Journal of Food Science*, **29**, 479–82.

Sumner, J. L., Samaraweera, I., Jayaweera, V. and Fonseka, G. (1982). *Journal of Food Science and Agriculture*, **33**, 802–8.

Taylor, J. A., Miller, D. C., Barrow, G. I., Cann, D. C. and Taylor, L. Y. (1982). In: *Isolation and Identification of Food Poisoning Organisms*. (Eds J. E. L. Corry, D. Roberts and F. A. Skinner), Academic Press, London.

Thomas, C. J. and McMeekin, T. A. (1980). *Journal of Applied Bacteriology*, **49**, 339–45.

Thomas, Y. O., Lulves, W. J. and Kraft, A. A. (1981). *Journal of Food Science*, **46**, 1951–2.

Thrasher, S. and Richardson, G. H. (1980). *Journal of Food Protection*, **43**(10), 763–4.

Trakulchang, S. P. and Kraft, A. A. (1977). *Journal of Food Science*, **42**, 518–21.

Vassiliadis, P. (1983). *Journal of Applied Bacteriology*, **54**(1), 69–77.

Vassiliadis, P., Kalapothaki, V., Trichopoulos, D., Mavrommati, C. and Serie, C. (1981). *Applied Environmental Microbiology*, **42**, 615–18.

Warseck, M., Ray, B. and Speck, M. L. (1973). *Applied Microbiology*, **26**(6), 919–24.

White, C. A. and Hall, L. P. (1984a). *Food Microbiology*, **1**, 29–39.

White, C. A. and Hall, L. P. (1984b). *Food Microbiology*, **1**, 97–104.

Index